ASPECTS OF TOURISM
Series Editors: Chris Cooper (*Oxford Brookes University, UK*), C. Michael Hall (*University of Canterbury, New Zealand*) and Dallen J. Timothy (*Arizona State University, USA*)

Polar Tourism

An Environmental Perspective

Bernard Stonehouse and John M. Snyder

CHANNEL VIEW PUBLICATIONS
Bristol • Buffalo • Toronto

Library of Congress Cataloging in Publication Data
A catalog record for this book is available from the Library of Congress.
Stonehouse, Bernard
Polar Tourism: An Environmental Perspective/Bernard Stonehouse and John Snyder.
Aspects of Tourism: 43
Includes bibliographical references.
1. Tourism–Polar regions. 2. Tourism--Environmental aspects–Polar regions. 3. Polar regions–Description and travel. 4. Polar regions–Environmental conditions. I. Snyder, J. (John), 1946-
II. Title. III. Series.
G155.P65S76 2010
910.911–dc22 2010005010

British Library Cataloguing in Publication Data
A catalogue entry for this book is available from the British Library.

ISBN-13: 978-1-84541-146-6 (hbk)
ISBN-13: 978-1-84541-145-9 (pbk)

Channel View Publications
UK: St Nicholas House, 31-34 High Street, Bristol BS1 2AW, UK.
USA: UTP, 2250 Military Road, Tonawanda, NY 14150, USA.
Canada: UTP, 5201 Dufferin Street, North York, Ontario M3H 5T8, Canada.

The policy of Multilingual Matters/Channel View Publications is to use papers that are natural, renewable and recyclable products, made from wood grown in sustainable forests. In the manufacturing process of our books, and to further support our policy, preference is given to printers that have FSC and PEFC Chain of Custody certification. The FSC and/or PEFC logos will appear on those books where full certification has been granted to the printer concerned.

Typeset by Datapage International Ltd.
Printed and bound in Great Britain by Short Run Press Ltd.

Contents

Authors' Introduction

'Tourism' can be defined in various ways. In the simple terms of the 1993 Chambers Dictionary, a tourist is 'a person who travels for pleasure', and 'tourism' is 'the activities of tourists and those who cater for them, especially when regarded as an industry.' Polar tourism, then, is the branch of an industry that caters for people who travel for pleasure to polar regions.

Do many people travel for pleasure to polar regions? Aren't they those bitterly cold regions at the ends of the earth – who could possibly seek pleasure there? The answer to both is 'yes': they are generally cold, but plenty of people want to visit them. Tens and hundreds of thousands per year are already visiting, with enough extra year-on-year to make polar tourism one of the fastest-growing branches of the worldwide tourism industry.

It was not always so: polar regions were not among the earliest tourism venues. The Arctic, and even more the Antarctic, stood well away from the trade routes of the post-mediaeval world. To reach them in the 18th and early 19th centuries, when tourism began, was seldom undertaken for pleasure or recreation. When travel conditions started to improve, these regions were hardly known to exist, and anyway, there were far more attractive and promising places in temperate and tropical regions to be visited first.

The Arctic was the first to be explored, by whalers, sealers, trappers and gold-miners pressing north to seek fortunes at sea and on land. Some of them wrote of its isolation, majestic scenery and abundant wildlife, prompting a few mid-19th-century entrepreneurs to set up what nowadays we would call small-scale 'nature tourism' or 'adventure travel'. The Arctic was but a few days' travel by steamer – later by rail or road – from northern centres of civilisation, so could be reached fairly easily by those who had money, time and a sense of adventure.

The Antarctic, discovered several generations later, lay much further from civilisation and retained its isolation longer. Outlined by maritime explorers in the mid-19th century, it was discovered and explored mostly in the early 20th century, and found to be similar to the Arctic but different – similar in cold, remoteness, scenic beauty and wealth of wildlife, different in being based around a continent rather than an ocean basin, and harder to reach through protective bastions of pack ice.

Several more generations passed before the Antarctic too became a venue for tourism.

Similarities have made both polar regions attractive to the same kinds of tourists – those who relish nature and wildlife in out-of-the-way places. Differences have led to the predominance of quite different forms of tourism in the two regions, with a wider spectrum of activities possible in the north than in the south. Differences too have arisen from the different kinds of regulation that have evolved, under national sovereignties in the Arctic, and international stewardship in the Antarctic.

This book is intended for readers with interests in how tourism develops in remote places, how the industry works in practice and how polar tourism is regulated and managed under the different political regimes that exist north and south. Its writers are an environmental biologist (BS), and a management consultant (JMS), both with considerable practical experience of tourism in polar regions. Its main concern is how regulation and management for the protection of environmental resources and preservation of human life can be accomplished in harsh marine and wilderness settings, where support services are scarce or non-existent. Tourism is growing rapidly in both polar regions, emphasising the need for positive responses to these management challenges. Maps of both regions can be found in Appendix F (p. 207).

Though centred on polar environments, our book raises issues and arguments that apply in principle to other forms of tourism in other sensitive areas of the world. We hope it may prove useful as a sourcebook, in particular to students of tourism management who are becoming aware of the environmental problems that tourism generates, wherever it occurs. Environmental issues are generally simpler in polar regions, making them good starting points for study.

Regulation of tourism is the business of authorities, whether local, national or (in the case of Antarctica) international. Management of tourism is the business of tour operators and authorities, in different but overlapping fields. Both have the dual responsibilities of preventing harm to both the environment and the tourists. Sensitive environments must be protected from the damage that tourism can readily inflict. And tourists travelling in wilderness regions must be safely conveyed. Planning and exercising forethought are required by tour operators and governments. They are most adept in their protective rôle when they can plan together, each minding its business in harmony with the other. Are polar regions well served by tourism's regulators and managers? Could they be better served by more efficient planning, clearer regulations and firmer management? That is what this book is about.

Bernard Stonehouse
John M. Snyder

Chapter 1

Arctic and Antarctic: Polar Regions and Environments

Introducing the Polar Regions

Polar regions are the areas of land and ocean surrounding the north and south geographic poles: for maps see the inside covers of this book. Though most people think of them as very similar – essentially cold, hostile and isolated from the rest of the world – geographically they are very different:

- The Arctic region, in the far north, is centred on a sea basin, partly ringed by lands that are mostly poleward extensions of the northern continents.
- The Antarctic region, in the far south, is centred on a large ice-covered continent, isolated from other continents by a wide expanse of ocean.

Both regions lie far from historic centres of civilization, and for centuries were regarded as too remote, difficult and unproductive to be worth exploring. The spread of commercialism during the 18th and 19th centuries saw the maritime fringes of both regions explored, mainly for possible access to China via northeast, northwest and southern passages. The northern seaways proved elusive, but their approaches were quickly exploited for walrus ivory, seal skins, seal and whale oil and baleen (whale bone). Southern islands too were stripped of their fur seals and elephant seals, and Arctic lands in both New and Old Worlds were pillaged for furs. Late 19th- and early 20th-century explorers extended geographical knowledge in both regions, mapping, prospecting for minerals, investigating the human communities, studying influences of polar regions on the rest of the world, ultimately reaching both the north and south geographic poles.

Both geographic poles were first visited by man during the early 20th century, and both can now be visited by tourists. Visitors to the North Pole are likely to arrive by Russian icebreaker (Figure 1.1) on scheduled summer cruises from Murmansk. The ship's Global Position System (GPS) indicator will tell them when they have arrived. Once down the gangway they will be standing at sea level on a sheet of ice 3–5 m thick. Air temperature will be a little below freezing point. The ice is drifting slowly toward Greenland and Svalbard. There is no permanent habitation, and unlikely to be anyone else around. The ship's company

Figure 1.1 *Yamal*, one of Russia's nuclear-powered icebreakers, was commissioned to keep open the Northeast Passage for cargo ships from Europe to the Far East. Capable of moving steadily through ice 10 m thick, *Yamal* now takes tourist passengers on three or four scheduled voyages each summer from Murmansk to the North Pole. Photo: BS.

set up a striped pole labelled 'NORTH POLE' and everyone dances around it. Hardy folk swim under careful supervision: the less hardy celebrate with Russian wine and a barbecue.

Visitors to the South Pole arrive overland by aircraft, tractor or on skis, probably with summer parties from coastal stations. They will be standing at around 2835 m above sea level, on a snow plain that overlies a cap of permanent ice some 3000 m thick. They will probably have GPS, but they'll also be in sight of a US scientific station that has been there since 1957. Air temperature will be around $-40°$. The position of the pole is marked by a barber's pole, ringed by oil drums and national flags. Station personnel may come out to meet them or invite them to visit, but not for long: there is work to be done.

Little now remains to be discovered geographically in either polar region, but both are still subject to intense scientific research in disciplines ranging from anthropology to zoology. In the mid-to-late 20th century, the Arctic became the front line in the struggle for ascendency between the Soviet Union and the USA. Called the 'Cold

War', it involved both contestants and their allies menacing each other across the Arctic Ocean and establishing defensive positions on Arctic lands. By contrast, the Antarctic became a zone devoted to peace and the pursuit of science under the international Antarctic Treaty.

In the late 20th century, both regions became popular tourist venues, and both are rapidly increasing in public prominence as polar tourism develops and proliferates. This is surprising, because neither has shaken off its image as a stomping-ground for explorers and eccentrics. Despite global warming, both regions remain predominantly cold and windy, uncomfortable and far from user-friendly. Precipitation falls mainly as snow or sleet. Where the snow settles and persists, it compacts under its own weight to form ice. Liquid water is scarce: in the colder areas, lakes, ponds, streams, ground surfaces and soils remain frozen for much of the year. Surface waters of the sea freeze annually to a depth of a metre or more, producing fast ice and pack ice (pp. 8, 17) that inhibit navigation for several months each year – almost entirely in winter, and to varying degrees in summer.

Latitude for latitude the Arctic region is warmer than the Antarctic. Arctic lands are generally low lying, their climates dominated by the sea at temperatures close to freezing point. The Antarctic's high, central continent reaches much lower temperatures, providing a vast heat sink that chills the whole region and much of the southern hemisphere beyond.

Geographically, the Arctic includes or abuts on the sub-polar northern extremities of Asia, Europe and North America – continents from which it has for millennia recruited animals and plants hardy enough to withstand its extreme conditions. By contrast, Antarctica is separated from South America, its nearest neighbouring continent, by almost 1000 km (over 600 miles) of ocean, and stands in ecological isolation from the rest of the world. Its very limited terrestrial flora and fauna are recruited mainly from species that have been blown there and subsequently adapted to its harsher environments.

Despite their reputations as areas of cold and hard living conditions, both polar regions have proved attractive to tourists – the Arctic for over two centuries, the Antarctic to a more limited extent for over half a century. How tourism developed in both is discussed in Chapters 2 and 3.

Polar boundaries

Though both polar regions are well recognised and often cited in scientific and common usage, there is no general agreement on their geographical limits. This becomes important in certain contexts where precision is needed: in studies of tourism, for example, different authorities may use or imply different criteria for defining 'polar' in their published statistics of regional tourism. We provide no single answer – just a warning that 'polar' does not in itself imply a single precise boundary.

- *Geographers* traditionally divide the world into latitudinal zones bounded by particular parallels of latitude. For polar regions the selected boundaries are the Arctic and Antarctic circles, respectively, 66° 32′ north and south of the equator. Either circle lies at the same distance of 2606 km (1619 miles) from its pole, and includes 40,333,466 km^2 (15,755,260 sq miles), roughly 8% of Earth's surface. The circles' angular position, defined by the angle of Earth's axis to the plane of Earth's rotation about the sun, alters slightly from year to year within limits of a few minutes of arc (Figure 1.2). Beyond either circle the sun remains above the horizon for 24 hours on midsummer day, and below it for 24 hours on midwinter day.
- Polar circles, like other parallels of latitude, make good, clearly defined boundaries that can readily be shown on maps – characteristics that recommend them particularly to lawyers and legislators. However, they do not show up on Earth's surface, and do not separate polar from non-polar climates or ecologies. The areas they circumscribe differ widely in content. The Arctic Circle includes forests, farmlands, towns, cities, industries and a human population of 4 million. The Antarctic Circle rings a desert continent without trees, shrubs or continuous ground cover, with a transient human population numbered in hundreds. Anyone concerned with climatic, ecological or socio-political interests looks beyond the circles for more meaningful regional boundaries.
- *Climatologists* define polar regions as the circumpolar areas in which the mean temperature of the warmest month does not exceed 10°C (50°F), i.e. where summers are always cool enough to define a distinctive climatic zone. Such areas lie within a boundary provided by the 10°C summer isotherm north and south. This boundary indicates more clearly than polar circles the areas within which living conditions remain cold for plants, animals and people even during the warmest season, and are thus limiting for many life processes. However, it provides no information on climate in other seasons, so it has no value in defining year-round polar conditions.
- *Ecologists* look for boundaries that separate recognisable 'polar' from 'sub-polar' plant and animal communities. They find two, respectively, for land and oceans. In the north, the treeline, or poleward limit beyond which forests are replaced by tundra or polar desert, is a useful terrestrial ecological boundary. It is

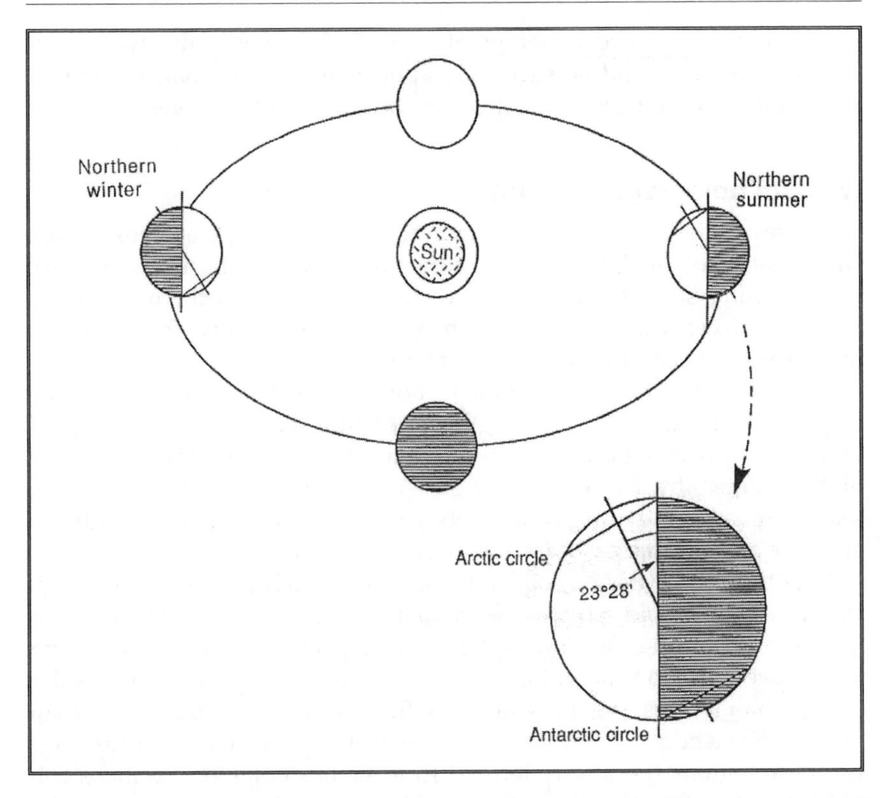

Figure 1.2 Earth's axis of rotation is tilted about 23°28′ from the plane of Earth's rotation about the sun. Points poleward of the polar circles thus receive different amounts of sunlight according to season. Observers in the Arctic see no direct sun in the northern winter, while for those in the Antarctic, the sun is above the horizon for 24 hours.
Source: Stonehouse (1989)

circumpolar where land exists, visible on the ground and in aerial photographs, and easily represented on maps. The southern hemisphere has no direct equivalent. There is a recognisable treeline only at the tip of South America: a notional one passes well to the south of the other southern continents.

- Instead, southern ecologists use the maritime Antarctic Convergence or Polar Front, a variable zone on the ocean surface, usually a few kilometres wide, marking the boundary between polar and sub-polar water masses. There cold polar surface waters, in summer diluted by melting sea ice, meet and disappear below warmer, more saline waters from sub-polar regions. The boundary is sometimes visible at the surface in calm conditions. More usually it is detectable by a sudden difference in temperature of 2–3°C where the two water masses meet, and as such may appear in satellite imagery. Similar boundaries appear in sea areas of the northern hemisphere.

Why are polar regions cold?

However defined, and by whatever boundaries, polar regions are consistently colder than the rest of the world. This is basically because the incoming solar radiation that warms Earth strikes them obliquely rather than vertically. In consequence, they receive only about 40% as much incident radiation as equatorial regions.

Polar regions differ too from the rest of the world in their seasonal day lengths. In the tropics, day length varies little between summers and winters. In polar regions, summers are marked by long days and short nights, winters by the reverse. At the geographic poles, day length and seasons merge: the sun remains above the horizon for six months in summer and disappears for six months in winter.

Like the rest of Earth, polar regions radiate energy to space through the year. They would radiate considerably more but for the blanketing or 'greenhouse' effect of the atmosphere, which keeps some of the returning long-wave radiation (heat) in. As a whole the Arctic receives more solar energy than it loses, the more highly reflective Antarctic loses more than it gains (Stonehouse, 1989: 33–34). Mean air temperatures that characterise these regions (i.e. those derived from values obtained by meteorologists at standard heights close to the ground) are maintained by poleward flows of heat, borne by atmospheric and oceanic currents. Thus, the temperature at any time and in any part of the Arctic or Antarctic represents a point of balance between radiation gains and losses, and incoming heat from surrounding areas, all of which remain relatively stable from year to year.

These balances are, however, subject to long-term variations over time-scales measurable in millennia, and shorter-term variations measurable in centuries or even decades, both due mainly to fluctuations in the amount of radiation received from the sun. Changes that occur near the boundaries of polar regions, which involve a shift between freezing and non-freezing conditions, tend to be self-sustaining. Decreasing radiation, for example, chills ground and air, which allows more snow and ice to settle and persist: the resulting whiteness increases the albedo or reflectivity of the surface, reflecting more solar radiation back into space and causing further chilling. Thus, polar climates tend to be less stable and subject to more frequent and more drastic changes than those of temperate and tropical regions.

The current frigidity of polar regions, marked by the presence of icecaps and both permanent and seasonal sea ice at both ends of Earth, is an anomaly in the long-term history of the world. Humanity has arisen during an ice age that has lasted some 20–25 million years, in which the icy conditions of both polar regions have alternately spread and diminished as a result of both long-term and short-term shifts in the balance of radiation and other controlling factors. As little as ten thousand years ago, icecaps spread far more widely across the northern hemisphere, covering huge areas of northern Europe, Asia and North America.

Since then, gradual warming has caused the ice to retreat to its present limits, notably in Greenland and the Canadian Arctic islands. The southern hemisphere has similarly warmed, reducing glaciers in southern South America and New Zealand and many of the islands fringing Antarctica, and to a lesser degree those fringing the continent itself. The retreat is continuing today, almost certainly accelerated by the consequences of human industrial activity (pp. 65–68).

Within this overall long-term trend, occupying several millennia, polar and sub-polar climates remain changeable in the shorter terms of centuries and decades. Short-term changes produce warm and cold spells that must have radically affected the distribution and welfare of plant and animal communities in pre-historic times, and have certainly exerted control over human activities during recent centuries of exploration, occupation and exploitation. Similar changes in future decades cannot fail to influence ongoing human activities, including the developing tourist industry (Chapter 4).

The Arctic Region

Geography

In this chapter, we define the Arctic region by its main ecological boundary, the treeline. Central to the region lies the Arctic Ocean, a rectangular basin 2500 km wide between Canada and Siberia, 4000 km long from Alaska to Norway, occupying in all some 14 million km^2. East of the Greenwich meridian its shores are Eurasian, fringed by the Norwegian, Barents, Kara, Laptev and East Siberian seas. West of the meridian lie Greenland, Canada, Alaska and the eastern extremity of Siberia, fringed by the Chukchi and Beaufort seas. The central basin, over 4000 m deep, is divided into two sub-basins by the submarine Lomonosov Ridge, which crosses almost directly under the North Pole.

Almost all of the Arctic Ocean is ice-covered in winter. Off the coasts, annual ice forms each autumn and persists until spring, reaching thicknesses of 1–2 m. The centre is occupied by a mass of perennial ice that fails to melt completely during the short intervening summers, and reaches thicknesses of 5–10 m. Annual ice also forms in neighbouring seas and channels, local conditions determining when it appears and for how many months it persists. In recent years, climatic warming has reduced the amounts of both annual and perennial ice, increasing the period in which ships can navigate Arctic waters, thereby stimulating trade in Arctic ports, but causing concern for wildlife, especially seals and polar bears, which may be affected (Chapter 4).

The Siberian coast is fringed by a continental shelf up to 1000 km wide and less than 100 m deep – a vast submarine plain totalling almost half the area of the ocean. From this coast flow five large rivers that pour fresh water into the Arctic basin: the Lena alone delivers over 500 km^3 annually, some 35% in June, helping to disperse the last of the inshore sea ice. Off north Greenland and the Canadian archipelago the shelf is less than 200 km wide, the continental slope descends rapidly to depths of over 3000 m, and there is little input of fresh water. Along the Siberian coast, fleets of Soviet icebreakers have traditionally kept a 'Northeast Passage' open for shipping through summer, providing an important trade route between Siberian ports. Off north Greenland and North America, sea ice presses against the coasts and fills the inter-island channels, making the 'Northwest Passage' more difficult to maintain than its Siberian counterpart. Warming has recently improved prospects for both passages. The Northwest Passage, formerly barred by ice in all but exceptionally warm seasons, is now traversed several times yearly by passenger cruise vessels.

Tourism in Arctic lands is outlined in Chapter 2. The Arctic presents a wide range of scenic attractions, including islands with steep fringing cliffs and headlands, alpine mountains, icecaps with rugged glaciers, fjords, beaches and broad coastal plains, lakes, rivers and spectacular waterfalls. Except close to settlements, they appear wild and untouched, and for management purposes most are indeed classed as wilderness (defined more fully in Chapter 5). Through a tourist season of three to five summer months, they remain predominantly green, and on sunny days appear welcoming alike to cruise ship visitors and to land-based travellers who prefer more active vacations.

- *Greenland*. Reputedly the world's largest island, it is in fact at least three islands part-hidden under an icecap that is over 3000 m thick in places. Though constantly renewed by snowfall, the cap has retreated along its east, south and western edges, exposing a picturesque and rugged terrain of mountains, fjords and tundra-covered coastal plains. In the far north, starved of snow, much of the ice has disappeared to expose bare ground and a patchy polar desert vegetation. The more hospitable western and southern fjords supported generations of nomadic Inuit tribes, living mainly by hunting on sea and land. From mediaeval times, the most sheltered and fertile areas were farmed by Scandinavian colonists, grazing introduced cattle and sheep, whose ruined settlements are now a tourist attraction. More recently, they supported Danish colonists, originally missionaries and traders, whose influence encouraged settlement and the development of villages and townships. Greenlanders – native, Danish and of mixed descent – still hunt and fish, but are now promoting commercial activities including tourism.

- *Iceland.* Geologically a new island, Iceland straddles the Mid-Atlantic Rift. It is consequently volcanically active, with hot springs, geysers and occasional eruptions that produce new coastal islets. An icecap with fringing glaciers occupies the southeastern corner: the centre is rugged tundra, grazed by sheep, cattle and horses, and drained by fast-flowing streams with picturesque waterfalls. Its predominantly Nordic population, descendants of early Viking settlers, enjoy a high standard of living based mainly on off-shore fishing. Tourism is rapidly developing and proliferating.
- *Svalbard.* This isolated, mountainous archipelago of five large and many smaller islands, ice-capped and remote in the northern Norwegian Sea, was for many years unclaimed politically, but exploited by many nations for coal and other minerals, furs, skins, walrus ivory and whales. In 1922, it became a province of Norway, and in recent years has developed a tourist industry based on readily accessible wilderness, enticing coastal scenery and huge populations of migrant land and sea birds.
- *Northern Scandinavia.* Arctic Norway, Sweden and Finland beyond the treeline benefit from warmth imparted by the North Atlantic Drift, which keeps their ports ice-free throughout the year, and provides tolerable living conditions far north of the Arctic Circle. In summer, they present well-managed wilderness tundra scenery, colourful fishing ports and high-quality tourist accommodation, equally accessible by cruise ship, flights or road. Human interest is provided by Saami herding and fishing communities. Rovaniemi, on the Arctic Circle in central Finland, is a centre of Saami culture, a university town and the authentic home and business centre of Santa Claus himself.
- *Northern Russia.* The isolated archipelagos of the polar basin, Novaya Zemlya, Zemlya Frontsa Iosifa (Franz Joseph Land), Severnaya Zemlya and the Noraya Sibir, New Siberian Islands, are mainly snow-covered and ringed by sea ice for much of the year. During short summers, their meagre tundra support huge populations of migrant birds. The long coast and hinterland of mainland Asia, low lying, tundra clad, and again snow-free for only a few summer weeks, is home to a range of Arctic folk, including the eastern Saami, Samoyeds, Nenets, Yakuts and Chukchi. These share

common ancestry with the Inuit, but base their economies more on land than on the sea – fishing in the estuaries and tributaries of the huge Siberian rivers, hunting, trapping and reindeer herding. The coastal plains support an extraordinary wealth of wildlife, now cherished both for its own sake and as a considerable attraction for wildlife tourism.

- *Northern Canada and Alaska*. The northern islands of the Bering Sea, the northern plains, glaciers and waterways of Arctic Alaska and Canada, the historically potent islands and channels of the Canadian Arctic archipelago – all are wilderness areas that formerly supported nomadic Inuit peoples along the coasts, mainly forest-dwelling 'Indians' inland and along the rivers. Though snow-covered in winter, in summer they are rich in immigrant birds and mammals, including huge herds of caribou. Valued originally by European settlers as a seemingly inexhaustible source of furs, these lands were among the first Arctic and sub-Arctic venues to be used for hunting, fishing and wildlife vacations. Improved transport, climatic amelioration and determined entrepreneurs combine to make them more accessible for tourism every year.

Climates and weather

Polar climates are characteristically cold, dry and windy, and travellers' tales have represented the Arctic as uncomfortably cold and dangerous throughout the year. However, within its vast expanses are many varieties of climate and weather. Tourists visiting the Arctic in summer, expecting blizzards or at least intense cold, are often surprised to find the air warm, little or no wind, and the sun intense enough to require screening. Visitors from the American or Canadian mid-west are even more surprised to learn that winter temperatures in their home towns often fall lower than in parts of Arctic Greenland, Iceland or Scandinavia. The coldest winter areas of the northern hemisphere lie not in the Arctic, but in the central heartlands of Siberia and North America.

Table 1.1 shows month-by-month temperatures for 10 Arctic localities. Their positions are given in Appendix A, together with temperature diagrams. Reykjavik, Iceland, has a characteristic maritime climate, in which the nearby sea, near-freezing but unencumbered by ice, chills the atmosphere in summer but warms it in winter. Its mean annual temperatures lie above freezing point, as do those of Angmagssalik (southeastern Greenland), Vardø (northern Norway), Arkhangel'sk (Russia) and Yakatut (southern Alaska). By contrast, stations in the Canadian archipelago, Whitehorse (southwestern Canada), Barrow (northern Alaska) and Verkhoyansk (eastern Siberia) have characteristic

Table 1.1 Monthly mean temperatures (°C) of 10 Arctic stations

	Months														
Arctic	*J*	*F*	*M*	*A*	*M*	*J*	*J*	*A*	*S*	*O*	*N*	*D*	*Mean*	*Range*	*Elevation (m)*
Kaujuitoq, Canada	– 31.8	– 33.7	– 31.4	– 22.1	– 10.2	0.6	4.6	2.9	– 4.4	– 14.6	– 24.9	– 29.3	– 16.2	38.3	64
Verkhoyansk, Russia	– 46.8	– 43.1	– 30.2	– 13.5	2.7	12.9	15.7	11.4	2.7	– 14.3	– 35.7	– 44.5	– 15.2	62.5	137
Barrow, Alaska	– 26.8	– 27.9	– 25.9	– 17.7	– 7.6	0.6	3.9	3.3	– 0.8	– 8.6	– 18.2	– 24.0	– 12.4	31.8	4
Isfjord, Svalbard	– 10.3	– 9.9	– 11.9	– 8.2	– 2.7	2.1	5.0	4.5	1.3	– 2.4	– 5.3	– 7.9	– 3.8	16.9	9
Angmagssalik, Greenland	– 7.5	– 7.8	– 6.4	– 3.5	1.4	4.9	6.6	6.6	4.1	0.4	– 2.8	– 5.7	– 0.8	14.4	35
Whitehorse, Canada	– 18.1	– 14.1	– 7.6	0.1	7.5	12.6	14.2	12.4	7.9	0.7	– 8.2	– 15.1	– 0.7	32.3	698
Archangel'sk, Russia	– 11.7	– 11.7	– 8.1	– 0.1	5.9	13.0	16.3	14.5	8.3	1.9	– 3.4	– 8.6	1.4	28.0	13

Table 1.1 *(Continued)*

Arctic	*Months*												*Mean*	*Range*	*Elevation (m)*
	J	*F*	*M*	*A*	*M*	*J*	*J*	*A*	*S*	*O*	*N*	*D*			
Vardø, Norway	−4.3	−5.2	−4.0	−0.8	2.6	6.2	9.1	9.7	6.8	2.5	−0.5	−2.7	1.6	14.9	15
Yakutat, Alaska	−2.6	−1.9	−0.3	2.8	7.0	10.3	12.3	12.1	9.6	5.5	1.0	−2.2	4.4	14.9	9
Reykjavik, Iceland	−0.4	−0.1	1.5	3.1	6.9	9.5	11.2	10.8	8.6	4.9	2.6	0.9	5.0	11.6	16

Source: Data from World Meteorological Organization (1971)
Note: For localities and temperature diagrams see Appendix A

continental climates, with fewer monthly means above freezing point and much lower annual means.

Arctic weather patterns will be familiar to visitors from lower latitudes who view weather reports on television, or follow radio bulletins for shipping. Much of the day-to-day weather is derived from successions of depressions (cyclones), or low-pressure areas, alternating with high-pressure areas or anticyclones, both tending to move from west to east at rates of 600–1000 km per day. Depressions spiral counter-clockwise, each bringing a recognisable sequence of lowering cloud, rain, sleet or snow in the warm sector, followed by a sharp change in wind and a spell of clear, cold, often showery weather. Between depressions come successive anticyclones – areas of clear, descending dry air, with internal clockwise circulation, which normally last for two or three days, occasionally longer. They provide clear skies, with bright sunshine during the day and often intense cold at night.

The mean course of depressions and anticyclones varies slightly from season to season, but overall they are responsible for most of the weather conditions experienced in the peripheral Arctic where polar tourism occurs. Within Arctic systems, the cold air tends to be colder, and the winds blow stronger, than in lower latitudes, with snow, sleet or hail more frequent than rain. Where mean monthly temperatures are lower, snow persists longer on the ground. Where annual mean temperatures are too low to melt the year's supply of snow, the ground becomes permanently covered and icecaps and glaciers dominate the landscape.

- In the central ocean basin, weather during the sunless winter is dominated by a stable anticyclone, with clear skies and light winds. The underlying mass of sea remains at temperatures close to freezing point, superficial sea ice providing a thermal blanket that allows only a little of its warmth to escape – enough to maintain air temperatures around −30°C. When the sun returns, the anticyclone weakens, allowing depressions to move into and across the basin. These bring warm air that partly melts the sea ice, and temperatures over the whole basin rise toward freezing point.
- Over North America and Siberia similar stable anticyclones form from November to May, bringing cold conditions over wide areas. Because there is no warming from an underlying sea, conditions are generally much colder than over the ocean basin. Verkhoyansk, close to the Arctic Circle in northern Siberia, holds the northern hemisphere record for extreme winter cold. Not quite so chill are central northern areas of North America and the Greenland icecap. Return of the sun in March or April weakens the anticyclone; warmer air and the sun's rays combine to melt winter snows and warm the ground, making June to August or September the

warmest months. Verkhoyansk's extreme summer-winter range of temperatures may well be a world record.

- Greenland supports an intermittent winter anticyclone, but its massive icecap results from heavy cyclonic snowfalls that may occur at any time of the year. The northeast coast is chilled by cold south-flowing currents that bring a year-round procession of sea ice from the Arctic basin. The west coast is warmed by northward-flowing currents, providing the relatively mild conditions that formerly encouraged Norse settlers to farm. Today, they bring cruise ships and visitors to many of the western ports.
- Iceland stands just south of the Arctic Circle, washed on its southern flank by the warm North Atlantic Drift, and on the north by the colder East Greenland Current. Sea ice often packed its northern shores during the 19th century, but is now almost unknown. Depressions sweep by throughout the year, bringing enough snow to maintain its small icecap and glaciers, and enough rain to dampen its thousands of enthusiastic tourists.
- Svalbard, Bjornøya and northern Norway also benefit from the North Atlantic Drift. Though all stand well north of the Arctic Circle, only Svalbard and its off-lying islands become icebound in winter, and all may count on receiving cruise ship passengers for three or four months every summer.

Vegetation and wildlife

Lands within the Arctic region are the northern extensions and off-lying islands of Europe, Asia and North America, from which they receive constant recruitment of plant and animal species, and from which came originally the ancestors of today's human populations (pp. 15–16). Though most of the land surface has been ice-covered during the past 10,000 years, close to sea level much of it has been ice-free long enough for soils to have formed and vegetation to establish itself. Northward from the treeline (or closely related 10°C summer isotherm), three vegetation zones can be distinguished:

- forest-tundra: a narrow transitional zone of stunted trees, shrubs and herbs;
- true tundra: a mixture of shrubs, herbs, grasses, reeds, mosses, liverworts, lichens and algae, low standing, with few plants above 1 m tall, many knee-high or prostrate;
- polar desert: a mix similar to tundra but thinner, with fewer species, mostly less than 30 cm high, thin soils and more bare ground.

Tundra, the vegetation that is most likely to be encountered by tourists, can take many forms. Depending on winter snow cover, drainage and other factors, it can be lush, with metre-high thickets of shrubs, grasses and other flowering plants, and dense mosses completely covering the ground, or thinner and drier, with plants growing only in favoured gullies that provide summer moisture and shelter from winds. It supports over a dozen major groups of invertebrates, including butterflies, bees, beetles, hover-flies, mosquitoes, springtails, spiders and mites, together with eight species of year-round resident land birds, 150 species of summer migrant birds and 40 species of mammals ranging in size from mice to musk-oxen.

Tundra in late spring and early summer has become a firm tourist attraction, often visited during landings from cruise ships, but also for nature-oriented tours planned expressly to see the flowers. Stonehouse writes:

> Early summer, shortly after the thaw, sees the tundra at its best and most colourful, with dwarf lupins, buttercups, windflowers, gentians and a host of other flowers making a spectacular carpet. From July onward grasses and shrubs begin to redden, and berries all too quickly colour the autumn scene. Summer walking on the southern tundra demands strong boots and insect repellents, for voracious mosquitoes and simuliid flies constantly seek a meal of blood to enhance their egg-laying. Birds and mammals are their usual victims, but human beings do just as well. (Stonehouse, 1990a: 92)

Arctic seas too are rich in wildlife, supporting huge concentrations of seabirds and an abundance of fish, seals and whales. Together with tundra wildlife, and the fish that seasonally migrate along the rivers, these have provided mainstays not only for the indigenous humans, but also for generations of commercial hunters, and more recently for parties of tourists on cruise ships and land expeditions armed with guns, cameras or both.

Man in the Arctic

The lands surrounding the Arctic Ocean form the northern extremities of eight sovereign states – Canada, Finland, Denmark (still partly responsible for Greenland and the Faroe Islands), Iceland, Norway, the Russian Federation, Sweden and the USA. The northern lands were settled originally by waves of mankind that invaded and successfully colonised over at least 3000 years. High Arctic indigenous people include the Inuit (formerly termed Eskimo), who inhabit the northern plains, coastlands and rivers of Greenland, North America and Asia. Arctic fringe peoples include the northern 'Indians' and Aleuts of North America, the Saami (Lapps) of Scandinavia and western Russia, and the Samoyeds, Yakuts and Chukchi of northern Siberia.

Starting over three centuries ago, the natural resources of the native people were exploited commercially by adventurers from the south, seeking furs and minerals on land, and seals, whales and fin fish from the sea. Exploitation became colonisation, and the northern peoples are now economically and socially integrated within the extended boundaries of the southern states. So an individual Saami may have Finnish, Norwegian, Swedish or Russian nationality; Inuit with many common features of ancestry live on both sides of the Bering Strait, northern Canada and Greenland. Interbreeding has occurred in all the stocks, and traditional cultures are diluted as the benefits of southern ways of life spread northward.

Each nation manages, develops and exploits its Arctic territories, and without exception sees tourism as a legitimate development, desirable in bringing revenues to areas that otherwise tend to drain resources. Impacts of tourism on indigenous populations are dealt with in Chapter 7.

The Antarctic Region

Geography

Here we define the Antarctic region by its ecological boundary, the Antarctic Convergence. So defined, it includes the central continent of Antarctica, divisible into two provinces, East and West Antarctica (the latter including Antarctic Peninsula) and the wide swathe of the Southern Ocean in which it lies. Close to the continent, south of South America, stands the chain of islands forming the Scotia Arc, including the South Orkney, South Shetland and South Sandwich Islands, and South Georgia. Elsewhere south of the Convergence lie a miscellany of other southern oceanic islands, including Bouvetøya, Heard Island, Iles Kerguelen, Macquarie Island and the so-called 'sub-Antarctic' islands of New Zealand, some of which receive regular visits from tourists.

Where questions of management arise, the Antarctic region has a second important defining boundary – the 60°S parallel of latitude. Although seven nations claim parts of Antarctica, within this boundary sovereignty and state ownership have, since 1961, been suspended under the terms of the international Antarctic Treaty. Management functions, which in the Arctic remain the responsibility of the eight sovereign powers, have been assumed by the international consortium of (currently) 48 nations responsible to the Treaty (pp. 150 and Appendix E). The 60°S parallel at all but a few points lies well south of the Convergence. Treaty rulings thus apply to management of all Antarctica, and the southern islands of the Scotia Arc. Still under sovereignty of national governments are South Georgia, the South Sandwich Islands and the remaining southern oceanic islands, here for convenience called the 'Southern Ocean islands'.

- *The Southern Ocean*. Encircling Antarctica like a moat, the Southern Ocean has an area of about 28 million km^2, flanked in the south by Antarctica and in the north by the Pacific, Atlantic and Indian oceans. Each winter, its southern half freezes over with ice up to 2 m thick, to an extent that can be monitored from satellites. From a March minimum of 3–5 million km^2, the total ice cover increases to a September maximum of 17–20 million km^2, investing the continent, the Peninsula and southern islands of the Scotia Arc.

  Late winter winds blowing off the land create inshore channels, and by November the pack ice has usually opened enough for ice-strengthened ships to reach and replenish coastal stations. Though much of the annual pack ice has melted by the end of summer, multi-year ice accumulates in local gyres, e.g. in the Weddell Sea, the central Ross Sea and at other points around the coast, providing local hazards to shipping that even the strongest icebreakers avoid.
- *Antarctica*. Fifth-largest of the world's continents, with an area of about 14 million km^2, Antarctica is roughly comma shaped and divisible into two provinces: the comma body is East Antarctica, the tail and its base form West Antarctica. The continent's mean surface elevation of over 2000 m is due mainly to the huge icecap, which covers all but 2% of the total area, leaving less than 5% of its coastline exposed, and rises in undulating stages to East Antarctica's central plateau at over 4200 m. The south geographic pole stands lower at 2835 m. Underlying topography, detected by radar from overflying aircraft, includes extensive mountain ranges, plains, lakes and islands, all completely obscured by ice.

  Geologically, East Antarctica is a remnant of the Gondwana supercontinent, formerly attached to Australia from which it split during the mid-Tertiary. Bearing much of the total icecap, its rocks are exposed mainly in the Transantarctic Mountains, which border the western Ross Sea in Victoria Land, south of New Zealand, and continue inland along the Ross Ice Shelf. There are more exposed mountains in Dronning Maud Land south of South Africa, MacRobertson Land south of Australia, and at intervals along the East Antarctica coastline. However, some 98% of its coast is lined with ice cliffs 20–40 m high, backed by extensive and featureless ice plains.

  West Antarctica, by contrast, is a southern continuation of the Andes – a range of ancient volcanic mountains and islands, the form of which is most apparent in the relatively exposed Antarctic Peninsula and neighbouring Scotia Arc. Further south, the rocks disappear almost completely under the icecap, though this province includes the continent's highest and most spectacular range, the Sentinel Range, with Vinson Massif (5140 m) its highest peak.

Antarctic Peninsula, the northernmost extension of the continent, is mountainous and heavily glaciated. Its east coast is permanently invested by the Weddell Sea gyre of multi-year ice. By contrast, the west coast is ice-free for three to four months each year, presenting a coastline of spectacular icefalls, fjords, rocky headlands and beaches. Snow-covered in winter, its beaches thaw out in summer, long enough to reveal a scant but intriguing vegetation of mosses and algae, alternating with extensive breeding colonies of penguins and other seabirds. Scenery and wildlife combine to make the Peninsula Antarctica's most popular area for tourism. How Antarctic and southern oceans tourism developed is dealt with in Chapter 3: how it is currently managed is the subject of Chapter 8.

- *The Scotia Arc.* The South Shetland, South Orkney and South Sandwich island groups form an extended, broken chain looping eastward and north from Antarctic Peninsula. The loop continues north, crossing the 60°S parallel south of the South Sandwich Islands, then turns westward to include South Georgia and ends in Staten Island and Tierra del Fuego. All the islands are of volcanic origin: the South Sandwich Islands are actively volcanic, and Deception Island in the South Shetlands has erupted intermittently during the two centuries that it has been known to man. All are mountainous and glaciated down to sea level, with indications of heavier glaciation in the recent past. The proximity of the South Shetland Islands to the 'gateway' ports of southern South America (p. 141) makes this group by far the most accessible point of contact between the Antarctic region and civilization. Its fjords and beaches provide harbourage and landing points for ships, with all the scenic and wildlife attractions of the Peninsula. The South Shetlands bear the highest concentrations of both scientific stations and summer visits by cruise ships. Further details of these islands appear in Chapter 8.
- *The peripheral southern islands.* Northern reaches of the Southern Ocean include islands that, lying south of the Antarctic Convergence, are included in the ecological Antarctic region, but because they lie north of 60°S, each remains under the sovereignty of a claimant nation. Southernmost are the South Sandwich Islands and South Georgia, forming part of the Scotia Arc (see above). Farther east lies tiny, isolated Bouvetøya, and beyond lie Heard Island, Iles Kerguelen and Macquarie Island, the latter two almost athwart the Convergence. South Georgia, the South Sandwich Islands, Bouvetøya and Heard Island are glaciated: South Georgia, Iles Kerguelen and Macquarie Island bear comparative rich vegetation of grasses and fellfield close to sea level, and are rich in stocks of seals and seabirds. Further details of these islands appear in Chapter 8.

Climates and weather

Like its northern equivalent, the Antarctic region has a reputation for a harsh climate and foul weather. This is based largely on the testimonies of early explorers, many of whom originated in temperate maritime climates and, not surprisingly, found polar conditions both dangerous and extremely uncomfortable. Again like the Arctic, Antarctica and its environs have many climates, which only year-round residents experience to the full.

Table 1.2 provides monthly mean temperatures for 10 Antarctic continental and fringe stations. Positions of the stations and temperature diagrams appear in Appendix A. Almost all continental stations have monthly and annual mean temperatures below freezing point. Exceptions are a few in the maritime Antarctic (Peninsula and Scotia Arc), in which summer means rise above freezing point, though annual means remain below.

Climatic zones can be characterised as follows:

- Antarctica itself has a strongly continental climate, dominated throughout the year by an almost permanently stable anticyclone over the icecap, and a constant succession of depressions moving west-to-east around the periphery. The central anticyclone brings extremely low temperatures, generally light winds and very little precipitation to the icecap. The depressions bring more variable weather to the periphery – warm, damp northwesterlies alternating with colder, drier winds from the continental slopes. Local katabatic (downslope) winds predominate in some coastal areas, for example in Terre Adélie and George V Land, where hurricanes are experienced almost daily.
- Antarctic Peninsula and the southern islands of the Scotia Arc, though invested in pack ice throughout winter, share a maritime Antarctic climate (p. 22), with warmer summers and milder winters than the rest of the continent. Summer weather, however, is often gloomy, with cloud and mist hiding the mountains for days on end. Ship-borne passengers on a ten-day cruise may consider themselves fortunate if two or three days are clear and sunny.
- South Georgia and the remaining peripheral islands, standing further from continental influences, have little or no winter sea ice. Depressions bring week-long persistent westerly winds with rain, snow, sleet and hail, punctuated by days of flat calm and brilliant sunshine. For further details see Chapter 8.

Summer cruise passengers visiting Antarctic Peninsula and the Scotia Arc, prepared for the worst with expensive foul-weather gear, are often surprised to find themselves enjoying warm sunshine on deck, dressed

Table 1.2 Monthly mean temperatures (°C) of 10 Antarctic stations

Antarctic	*Months*												*Mean*	*Range*	*Elevation (m)*
	J	*F*	*M*	*A*	*M*	*J*	*J*	*A*	*S*	*O*	*N*	*D*			
South Pole Station	– 27.9	– 40.2	– 54.3	– 57.3	– 57.3	– 58.2	– 59.9	– 59.7	– 58.4	– 50.7	– 38.4	– 27.7	– 49.3	32.2	2835
McMurdo Station	– 3.1	– 8.8	– 17.6	– 21.1	– 23.3	– 23.5	– 25.8	– 26.9	– 25.0	– 19.5	– 9.9	– 3.8	– 17.4	23.8	24
Mawson Station	0.1	– 4.4	– 10.3	– 14.5	– 16.1	– 16.8	– 17.8	– 18.8	– 17.7	– 13.2	– 5.4	– 0.3	– 11.3	18.9	8
Hope Bay	0.2	– 1.3	– 3.7	– 7.5	– 9.3	– 11.4	– 11.8	– 10.6	– 7.4	– 4.1	– 2.1	– 0.2	– 5.8	12.0	13
Faraday Station	– 0.1	– 0.3	– 1.4	– 4.9	– 7.7	– 10	– 12.8	– 11	– 8.6	– 4.9	– 2.4	– 0.7	– 5.4	12.7	10
Orcadas Station	0.3	0.5	– 0.6	– 3.0	– 6.7	– 9.8	– 10.5	– 9.8	– 6.4	– 3.4	– 2.1	– 0.5	– 4.3	11.0	4
Deception Island	1.4	1.1	0.1	– 2.1	– 4.3	– 6.3	– 8.0	– 7.7	– 4.8	– 2.4	– 1.0	0–5	– 2.8	9.4	8

Table 1.2 (*Continued*)

	Months														
Antarctic	*J*	*F*	*M*	*A*	*M*	*J*	*J*	*A*	*S*	*O*	*N*	*D*	*Mean*	*Range*	*Elevation (m)*
Grytviken	4.7	5.4	4.6	2.5	0.2	– 1.5	– 1.5	– 1.5	0.1	1.7	3.0	3.8	1.8	6.9	3
Macquarie Island	6.8	6.7	6.2	5.1	4.2	3.3	3.2	3.2	3.4	3.8	4.5	5.9	4.7	3.6	30
Stanley, Falkland Islands	8.7	9.0	8.2	5.8	3.9	2.4	2.2	2.6	3.4	5.2	7.0	7.7	5.5	6.8	12

Source: Data from World Meteorological Organization (1971)
Note: For localities and temperature diagrams see Appendix A

in little more than they would wear on a summer day back home. However, this works only if the ship is stationary or has a following wind approaching its own speed. Even a light wind can cause immediate chilling and discomfort. Weather too is notoriously changeable: a calm, clear morning with light winds can be followed by an overcast, blizzard-ladened afternoon. Passengers going ashore even for half an hour are advised always to carry spare jumpers, windproof outer clothing, gloves and a head-covering to reduce heat loss if the weather turns foul.

Vegetation and wildlife

Unlike the Arctic, Antarctica and its neighbouring islands are isolated from all other continents by a broad span of ocean. Recruitment from other continents is thus limited to plant propagules and animals that can fly, swim or be carried or blown there. Again unlike the Arctic, both continent and islands are still in the grip of an ice age: very little of their surface has been ice-free long enough to develop fertile soils, and such vegetation as exists is closer to Arctic desert than to even the poorest tundra. Only about 2% of continental Antarctica is ice-free, supporting small, widely scattered patches of mosses, lichens and algae.

Only marginally richer are Antarctic Peninsula and southern islands of the Scotia Arc (together forming the Maritime Antarctic), where slightly longer summers and damper conditions provide for dense moss beds, and patches of the region's two species of flowering plants, a hairgrass, *Deschampsia antarctica*, and a pearlwort, *Colobanthus quitensis*. Northern islands of the Scotia Arc, notably South Georgia, at sea level have a richer fellfield flora of grasses, ferns and small shrubs, prolific enough to support stocks of introduced reindeer. Neither the continent nor the islands support indigenous land mammals.

A major tourist attraction along Antarctic coasts is the massive populations of breeding seabirds, notably petrels and penguins, that line many of the beaches and cliffs, and the seals and whales that, after fierce 19th- and 20th-century predation by commercial hunters, are now restocking the southern oceans. In the absence of indigenous human populations south of the southern tip of South America, Antarctica's only human settlements are groups of scientists, whose homes, cultures and ways of life are closed to enquiring visitors except by special invitation.

Man in the Antarctic

Antarctica has no indigenous human population. Though boats carrying native people from South America or New Zealand may have been blown south toward the Antarctic region, there is no evidence of early human occupation or settlement. The earliest explorers of the southern oceans were naval expeditions and sealers of the 18th and early 19th centuries: the sealers worked on the outlying islands, where some may have spent involuntary winters. Thus, there are no Antarctic states equivalent to the eight Arctic states, contiguous with the continent and with resident human settlements.

The earliest residents who stayed longer than a few months were early 20th-century explorers and scientists overwintering in temporary research stations. From the mid-century onward, their successors occupied more permanent stations, mainly for scientific purposes, and to establish occupancy in support of political claims to sovereignty. Beltramino (1993) provides an account of human populations of the region during this period.

However, seven nations – Argentina, Australia, Chile, France, New Zealand, Norway and the UK – claim sectors of continental Antarctica and the adjacent islands (see endpage map and Table 8.1, p. 144). Their claims, adjacent or overlapping, leave a large sector of West Antarctica unclaimed – the area between 90°W and 150°W that is particularly difficult to reach by air or sea. Though some were based on 18th- or 19th-century discoveries, all were renewed or restated during the first half of the 20th century. The stimulus was the start of Antarctic whaling in 1904, which provided opportunities for taxation and revenues (Hart, 2001, 2006). Three of the claims (those of Argentina, Chile and the UK) overlap substantially, provoking disagreements that during the 1950s came close to hostilities. For a fuller discussion of Antarctic geopolitics see Dodds (1997).

The seven claims to Antarctica are not recognised by most other nations, mainly on the grounds that, up to the start of whaling and its subsequent spread, none of the claimants made serious efforts to occupy or administer the territories – a necessity by international convention. Most of the non-claimants show little concern for who claims to own Antarctica. Those with interests in the continent tend to be satisfied by accession to the Antarctic Treaty, which dissociates itself from sovereignty issues. Some non-claimants that have been involved in exploration take a more positive view. Russia and the USA, for example, do not respect existing claims, make no claims of their own, but reserve the right to claim parts of the continent in the future. As Triggs (1987b: 53) pointed out, 'It is ironic that these states far outstrip some of the claimants in present scientific and exploratory activities, and would arguably have better legal title in Antarctica were they to claim it'.

Claimant nations do not manage tourism within their claimed territories. For most practical purposes, responsibility for tourism management has been delegated entirely to the Antarctic Treaty System, which does not share the enthusiasm of the Arctic nations for tourism within its domain. For its regulating strategies see Chapters 8 and 9.

Summary and Conclusions

This chapter outlines similarities and differences between Earth's two polar regions. Both are remote from inhabited temperate and tropical regions, and both are colder at sea level than much of the rest of the world. Geographically, they differ widely: the Arctic is centred on a complex sea basin almost entirely ringed by land, the Antarctic on a high, ice-covered continent surrounded by deep ocean. Geographic, climatological and ecological boundaries are defined. For most environmental purposes, the treeline or 10°C mean summer isotherm delimits the north, the Antarctic Convergence (or polar front) the south. The political Antarctic region – the region controlled by the Antarctic Treaty System – is limited by latitude 60°S.

The prevailing cold and seasonality of polar regions are outlined, and the inconstancy of polar cold in world history is discussed. Though polar regions are always colder than the rest of the world, their present extreme frigidity indicates that Earth is currently in an ice age, manifest in the presence of polar icecaps. However, mean temperatures fluctuate, more so in the Arctic than in the relatively stable Antarctic, as shown by the retreat over the last 10,000 years of ice from both ends of Earth. Warming of polar and neighbouring temperate regions at present appears to be accelerating, almost certainly due to the increasing efficiency of the atmospheric blanket, due in turn to industrial gases and particles released by man into the atmosphere during the past few centuries.

The regions are compared in terms of geography, climates, wildlife and people, and with particular reference to their growing importance as venues for tourism.

Chapter 2
Arctic Tourism: History and Development

Introduction: Unlikely Regions for Tourism

Tourism is travel for pleasure, and pleasure is not the first image that polar regions bring to the public mind. Remote, cold and inhospitable, yes: everyone knows the polar regions are far from civilisation, at the frozen ends of the world, and uncomfortable to live in (see Chapter 1). So who goes there on vacation?

Who goes there at all? Four million people from all walks of life live permanently in the Arctic, including indigenous folk, and entrepreneurs, administrators, scientists and many others, all concerned with using the region's natural resources simply for living, or to exploit them for profit in one field or another. To these must now be added millions of tourist visitors to the Arctic, who seasonally and locally consistently outnumber their host populations. The vast majority of visitors are quite ordinary folk who travel aboard mass transport, primarily cruise ships of all sizes, to experience a pleasurable journey. A substantial minority include persons seeking adventurous experiences, such as mountaineering, rafting, sport fishing, wildlife photography and hunting.

And who goes to the Antarctic? There are no indigenous or permanent populations – just a few hundred scientists and support staff who manage a few dozen research stations. But still, tens of thousands of tourists visit the Antarctic region annually, again mostly on cruise ships, again with a minority seeking more adventurous alternatives.

Tourism is the world's biggest industry, and polar tourism is one of its fastest-growing sectors. Travel for pleasure has now become the single largest human activity in both polar regions. Tourism to the Arctic began two centuries ago, grew slowly during its first century, and now each Arctic nation caters to hundreds of thousands of clients every year. Antarctic tourism began only 50 years ago, grew slowly during its first three decades, then picked up momentum: it now attracts tens of thousands per year. In the first decade of the 21st century, both are booming – growing, diversifying and setting challenges for public authorities whose business is to manage resources at either end of the world. This chapter outlines the history and development of tourism in the Arctic. For the development of tourism in the Antarctic region see Chapter 3.

How Arctic Tourism Developed

Arctic environmental features include an ocean with a surface frozen for much of the year, intense cold and abundant snow on land, and vast expanses of plains and mountains with underlying permafrost. Continental land masses and islands, governed by eight sovereign nations, encircle the ocean that, for most of human history, was virtually impenetrable. Indigenous human occupation of the region required a nomadic lifestyle, and the ability to tolerate severe weather and long nights enduring for nearly half the year.

During the warmer months, both land and marine birds and mammals migrate to the Arctic, where tundra vegetation is, for a short time, abundant and oceanic surface waters are prolific (Stonehouse, 1990: 87–89, 136–137). Over millennia, the native people who were its first permanent inhabitants developed lifestyles and technologies that enabled them to survive these extreme conditions.

Attempts to explore and occupy high latitudes to obtain economic benefits and expand empires from 1576 onward were beset by peril and tragedy (McGhee, 2001; Delgado, 1999). All expeditions experienced hardships, and many lost ships and men. By the early 1800s, newspaper articles and books describing both the heroic and tragic aspects of polar exploits had become popular reading. Remarkably, that is in fact when tourists began visiting the Arctic, and the attractions of this unlikely destination have grown steadily for more than two centuries, establishing a significant human and economic presence in the Arctic.

Pioneers of Arctic tourism

Snyder and Stonehouse (2007: Chapter 1) outlines the history of Arctic tourism. Its pioneers were 'knapsack' adventurers – independent mountaineers, sport anglers and hunters wealthy enough to pursue the recreational opportunities provided by wilderness, unique cultures, unexplored mountains and seasonally abundant wildlife. By 1807, a guidebook was available for travellers to the Scandinavian Arctic (Viken & Jørgensen, 1998). Most early tourists hiked or used local transport, and the generous hospitality of Arctic people enabled them to stay in local homes or camps, adding much to their experience.

By the mid-1800s, mountain climbing for recreation had come into vogue (Dent, 1892). Mountaineers made exploratory trips to Norway, and by the end of the century to Canada, Alaska, Spitsbergen and Greenland. Routes to summits and expedition advice for Arctic destinations were widely publicised in both individual accounts and the journals of newly founded alpine clubs (Forbes, 1853; Conway, 1897; Williams, 1859).

Arctic sport hunting and fishing, which also appeared in the early 19th century, were originally the sole province of the wealthy. Immense quantities of fish, especially salmon and trout, attracted anglers, and exotic trophy wildlife species such as bears, moose, caribou, Dall sheep and mountain goats attracted the hunters. Despite the disparity of wealth, remarkable partnerships evolved between the indigenous people of the Arctic and their sportsmen visitors. By the end of the 19th century, both an economic interdependency and cultural tolerance were well established between these disparate groups. Arctic sport fishing and hunting tourism, and the value of indigenous guides, were extensively chronicled in such recreation publications as *Field and Stream Magazine,* which were established at the time.

Tourism for the masses

By the 1850s, the Industrial Revolution in North America and Europe had created and spread personal wealth, increased leisure, improved access to education and generated new inventions and technologies that transformed societies. One result was an extraordinary expansion of tourism worldwide. Widespread wealth bought affordable passages on railroads and steamships: by the late 1800s, tourism had become viable for the masses, rather than the privileged few. Intense competition between railroad and steamship companies progressively lowered travel costs, allowing more people to travel to more destinations (Figure 2.1).

Figure 2.1 By 1880 the Arctic was a popular mass tourism destination. Photo: JMS collection.

Transport companies expanded their networks to previously inaccessible regions, including the Arctic. In 1850, commercial steamship tourism was initiated in Norway. By the 1880s, Arctic tourism and travel by steamship had become a booming business, to destinations including Norway's fjords and North Cape, Spitsbergen, Greenland, Baffin Bay, Iceland, Alaska's Glacier Bay and gold rush sites as far north as Homer, and riverboat cruises in the Canadian Yukon. Tourist experience aboard the steamships was a mixture of exploration and luxury, led by Arctic explorers and naturalists. Little-known or recently discovered glaciers, bays, wildlife and indigenous communities attracted curious tourists. Shipboard life included lavish meals, orchestras, beauty parlours, barbershops, photography studios and lectures presented within library settings (Murray, 1983; Collis, 1890; Burroughs *et al.*, 1986; Kneeland, 1876).

Within the same period, companies coordinated rail schedules with steamship departures, providing efficient travel for thousands of passengers. Railroad operators recognised the economic value of tourist destinations along their routes, and simultaneously both citizens and governments saw the value of preserving natural attractions as national parks and reserves. There arose remarkable partnerships between national governments and the railroads, governments designating national parks and monuments and encouraging their citizens to visit them, while railroads provided access, facilities and accommodation. National parks and monuments designated in Alaska extended this symbiosis to the Arctic: Canada established the world's first National Park System and Park Canada's elegant accommodations were built and operated by the railroads. Sweden established Europe's first national park, with access to this and subsequent parks provided by rail.

By 1900, Arctic tourism was a flourishing and diverse industry. Its diversity included independent travellers pursuing a variety of adventurous recreation activities in marine and land environments, as well as groups touring natural, wildlife, historical and cultural attractions. These activities were extensively promoted in guidebooks by John Murray, Baedeker and other specialist publishers (Dufferin, 1873; Kneeland, 1876; duChallier, 1881; Scidmore, 1885), successive editions regaling the splendours of the Land of the Midnight Sun. Literature encouraging mass travel regularly appeared in such widely distributed periodicals as *Harper's Weekly, Century Magazine* and the *National Geographic Society Magazine*.

Twentieth-century advances in transportation technologies, together with increased personal wealth, further propelled the growth and diversification of Arctic tourism. Commercial air transport, for example, not only expanded access, but also created new tourism experiences such as floatplane adventures and flight seeing. Other transport improvements such as four-wheel drive vehicles, snow machines and reliable small boat engines expanded the tourist's geographic range and provided opportunities to pursue independent travel. Simultaneously, economic competition arising from improved access and diversification of Arctic tourism products and services progressively made Arctic travel more affordable.

Economic benefits arising from Arctic tourism were immediately evident both to private companies and to Arctic governments. A departure from such resource-depleting industries as hydraulic mining, timber harvesting and the exploitive commercial fishing and whaling practices of the 19th and early 20th centuries, tourism represented a new way to use the Arctic's natural resources, and provided jobs, personal income, revenues and financial capital for infrastructure.

Arctic Tourism Today

Arctic tourism has evolved to become a mature and highly diversified industry that now operates throughout the year in all circumpolar nations (Figure 2.2). It plays a vital role in the economies of all circumpolar nations and Arctic people rely on the jobs, income and revenues it provides to sustain their lives. This interdependency must be acknowledged when determining how the industry will be managed and further developed.

Figure 2.2 Railways with vintage coaches transport tourists to remote Artic locations. Photo: JMS.

Unlike the Antarctic that has no resident population, the people of the Arctic require jobs and income to survive and tourism contributes significantly to their well-being.

Several major markets that comprise Arctic tourism are best defined in terms of their primary attractions and the ways in which those attractions are experienced.

- Mass tourism – seaborne and airborne.
- Fishing and hunting.
- Nature tourism.
- Adventure tourism.
- Culture and heritage tourism.

This approach to classifying tourist markets acknowledges tourist' expectations, the service delivery methods used to realise them and the probable impacts resulting from these activities. Each of the markets is growing and expanding for obvious reasons – they appeal to tourists who are willing to pay for the unique experiences they offer. All eight Arctic nations are aggressively seeking to develop tourism and this represents a new type of economic use for several immense Arctic regions. Most notably, the emergence of the Russian Federation as an Arctic tourism venue and the ambitious promotion of Greenland and Arctic Canada as tourist destinations have greatly expanded the economic, geographic and cultural dimensions of the industry.

Tourism now plays a vital role in all aspects of the Arctic world. The private and public economies of the circumpolar world are increasingly dependent on it for a variety of benefits. Native people seeking to participate in the market economy and reduce their reliance on scarce resources are especially attracted to tourism development. Natural and cultural resource management policies and plans affecting the Arctic's recreational resources define how those resources will be allowably and seasonally used. Local communities are seasonally transformed by growing numbers of tourists, and the cultural values of indigenous people are directly influenced by these contacts.

Seaborne mass tourism

This sector is comprised of tourists primarily attracted to sightseeing within the pleasurable surroundings of comfortable transport and accommodation. The cruise ship industry is the single largest provider of mass tourism in the Arctic. During the past few decades, its popularity has grown tremendously, as evidenced by increasing numbers of ships and passengers, increasing size of ships, growing economic importance for Arctic nations and geographic expansion to new venues. Such traditional cruise ship destinations as Norway and Alaska have witnessed rapid, continuous growth since the 1980s. Newer destinations, such as Greenland and Iceland, are becoming increasingly reliant on Arctic tourism and cruise ship traffic to sustain their economies.

Box 2.1 World cruising: A growing market

In 1990, the world cruise industry as a whole transported 4.5 million tourists to diverse international destinations. Between 1995 and 2005, worldwide demand for cruising more than doubled from 5.7 million to 14.4 million passengers (Figure 2.3). Over the same period, the number of Europeans taking cruise holidays around the world more than trebled from 1 million in 1995, to reach 3.4 million in 2006 (*World Maritime News*, 2008). Record cruise industry profits accompanied that growth: the most recent data indicate a $15.3 billion NET profit in 2004. The most conservative economic forecasts by cruise industry experts indicate that by 2010 at least 17 million passengers will travel by cruise ship (Brown, 2006; *Cruise Industry News*, 2004). Anticipated passenger growth is being matched by an ambitious ship-building programme that will last more than 15 years (Brown, 2005, 2006). For example, European shipyards are under contract to build 36 cruise ships with a combined value of €14.9 billion through to 2011 (*World Maritime News*, 2008). The industry is literally banking on increased cruise ship travel, of which Arctic cruising is likely to take its share.

Figure 2.3 Larger ships are entering the Arctic tourism market. Photo: JMS.

Arctic voyages are a vital and especially lucrative part of the industry's international product, constantly expanding in terms of ship passenger capacity, new destinations and extended seasons of operations. All the Arctic governments have become steadfast economic partners, supporting the cruise industry by means of infrastructure and community investments, tax incentives, marketing and promotion. In 2004, over 1.2 million passengers travelled to Arctic destinations aboard cruise ships. By 2007, the number had more than doubled. Equally impressive are the increased numbers of destinations, types of vessels serving this market, extended duration of the sailing season and diversity of shore excursion programmes offered by host countries. Examples of this remarkable growth and geographic expansion include the following:

- *North Pole.* Once the most formidable challenge of Arctic exploration, the North Pole is now routinely visited by cruise ships. Regular scheduling has replaced the infrequent visits that occurred between 1977 and 2004. Brigham and Ellis (2004) reported:

  > 'During 1977 to 2004, 52 successful voyages have been made to the North Pole by the icebreakers of Russia (42), Sweden (4), Germany (2), United States (2), Canada (1), and Norway (1); remarkably eight surface ships reached the North Pole during the summer of 2004. Thirteen of the voyages were in support of scientific research and the remaining 39 were devoted to tourist voyages to the North Pole and across the Arctic Ocean'.

  Six cruise ships were scheduled to sail there in 2008, with each passenger paying at least US$20,000.
- *Canadian Arctic.* The Northwest Passage through the Canadian Arctic was for the first time negotiated by a cruise ship, M/V *Lindblad Explorer* (Figure 2.4), in 1984. From then to 2004, a total of 23 commercial cruise ships and 15 recreational yachts completed the passage (Snyder & Shackleton, 1986; Headland, 2004), and cruise ship tours have since become annual events. The Canadian Ice Service reported that at least seven transits occurred in 2008. As sea ice continues to diminish, an increase in transits of the Northwest Passage are anticipated. Remarkable evidence of these changing conditions was reported by the Canadian Ice Service (2009), stating that 'for the first time in recorded history, national ice charts showed that both the Northwest Passage and the Northern Sea Route were simultaneously open water for a brief period in September 2008'. Cruises to other Canadian Arctic destinations doubled in 2006, from 11 to 22. In 2007 Inuit-owned Cruise North Expeditions announced the start of cruise ship operations. Their 2008 and 2009 cruises earned an excellent

Figure 2.4 *Lindblad Explorer* pioneered commercial tourism of the Northwest Passage. Photo: BS.

reputation for responsible operations and the sharing of Native culture (Stewart *et al.*, 2007; Canadian Press, 2007, http://www.cruisenorthexpeditions.com, 2010).

- *Norwegian Arctic.* Norway received 1.13 million cruise ship passengers in 2007, according to the European Cruise Council (*World Maritime News*, 2008), and numbers continue to grow. In 2007, international cruise lines added several new ports of call in Norway. Ålesund reported strong growth in 2007 resulting from 68 cruise calls, with a total of 60,000 passengers. Many cruises head north beyond Lofoten to Tromsø, North Cape and Svalbard. In 2007, Svalbard reported 45 cruise calls, 17 more than in 2006 (*Aftenposten*, 6/4/07). The size of vessels operating cruises in Norwegian waters also increased substantially in 2007, the first year that the major cruise line, Royal Caribbean, operated a vessel from Oslo, *Jewel of the Seas*; more use of Oslo as a cruise starting point is planned for 2008. Non-resident cruise ship tourists visiting Norway in 2004 (the most complete data available as of December 2006) spent 2.383 million NOK. Norwegian cruise expenditures have increased from 2.196 million NOK in 1998 (*Statistics Norway*, 2006a).
- *Greenland.* Greenland's cruise ship industry is growing dramatically. According to the Greenland Tourism Board, the number of cruise ship passengers visiting Greenland is increasing by about 30% annually. Between 2005 and 2008, cruise ship port calls increased from 56 to 375. The 375 port calls included 39 different ships with a combined total of 22,499 passengers. When the number of crew and staff are added to the number of passengers,

it is estimated that the total represents more than half of Greenland's 2006 population of 56,901. Substantial expansion of the cruise industry will occur when the Qaqortoq port development begins. Construction will enable large ships to berth in this southern region of Greenland and a collection of host facilities will be built to accommodate a variety of tourist needs (Greenland Tourism & Business Board, 2009; *Sermotsiaq*, 4/16/08).

- *Iceland.* Since 1990, Iceland's tourism has grown at an annual rate of at least 9%. Cruise ship passengers represent the fastest growing segment, particularly during 2003–2008. In 2003, a total of 320,000 foreign tourists visited Iceland and of that number, 31,200 were cruise ship passengers. By the summer of 2007, a total of 77 cruise ships made calls at nine Icelandic ports. In 2004, the 'Cruise Iceland' organisation was founded, which now includes 22 major cruise ship companies that are highly integrated with shore-based tour and transport companies. In 2008, the port of Reykjavik opened a large visitor and hospitality centre and a total of 12 ports welcomed cruise ships (www.icetourist.is, 2005; cruiseiceland.com, 2008).
- *Arctic Russia.* The Russian Federation transformed the Arctic cruise ship industry in two important ways. The first resulted in 1990 when Russian icebreakers and ice-strengthened vessels were used for commercial cruise operations. Nuclear-powered icebreakers and conventionally powered, ice-strengthened, Soviet-era vessels expanded the geographic range of Arctic tourism, extended the tourist season and increased passenger capacity. The most significant geographic expansion of Arctic tourism occurred when the Russian Federation allowed cruise ship entry to numerous Arctic seaways and ports throughout its vast domain. In January 2007, the Russian Ministry of Transport in coordination with regulations from the Ministry of Defence and the Federal Customs Service granted permission to open six Russian ports to foreign tourists. The first cruise ship passage through the Northern Sea Route was conducted in 2008. The Solovki Museum confirms that in 2008, for the first time in history, as many as six cruise liners will visit the Solovetski archipelago in the White Sea, carrying tourists of whom most will be American and German. 2008 will also initiate a partnership between Norway and Russia for tourist cruises in the White Sea and entry to destinations in Archangel'sk Oblast. Even Cold War installations, such as those in Chukotka, can now be visited with permission from the Russian government.

- *Alaska.* In 2004, Alaska received 876,000 cruise ship passengers from May through September. By the end of the 2007 season, 1,029,800 passengers had travelled aboard Alaska cruise ships. That single destination number is nearly equivalent to the total number of passengers who cruised the entire Arctic in 2004. Between 2006 and 2007, the number of Alaska passengers grew by 9.2%, and the number of cruise passengers was twice as large as the state's total population (Harpaz, 2005, State of Alaska, 2008). Based on 2004 rate schedules, passengers paid between $2000 and $20,000 per person for their cruises. In addition, each person spent an additional $82 per port visit (Cruise Industry News, 2004). As of 2008, the prices for Arctic cruises range between $2900 and an extravagant $55,000 per person.

The total economic value of cruise ship travel is calculated by including the money paid for the numerous shore excursion programmes that are offered, the purchase of goods and services, and land transport. Thus, shore-based travel by train and motor coach, land, air and sea excursions, and the purchase of products, services, accommodation and food contribute to the complete economic measure.

The preceding cruise ship information also reveals the growing reliance of native people on this form of tourism. In some instances, native people own and operate cruise vessels and in other settings they actively promote cruise ship entries to their ports and waters. The substantial investments in infrastructure, vessel operations and marketing promotion by native governments provide solid evidence that this form of tourism will have a long-term presence in the Arctic.

Airborne mass tourism

Modern air transport technologies, new airports and flight schedules have made the Arctic a more accessible place. Large commercial jets with enormous passenger capacities introduced a new form of mass tourism to the Arctic. By reducing both travel time and cost, they improved the competitive position of Arctic tourism, its attractiveness to potential tourists and the amount of time that the tourist could actually spend in the Arctic instead of travelling to it. All these improvements especially benefited the cruise ship industry and local economies. Increasing numbers of Arctic-bound tourists arrive by plane at airports located near embarkation ports, sailing itineraries gain more flexibility in terms of both routing and scheduling, and a greater diversity of shore excursion programmes are offered at more communities. Air operators and cruise ship companies have established coordinated schedule times, modes of transport that closely match passenger capacities, travel pricing arrangements and tourist marketing programmes.

All major cities and national capitals in the Arctic are now served by regularly scheduled commercial air services. Year-round access provides tourism gateways to the entire region and has expanded tourist seasons. For example, previously remote Arctic destinations, such as Iqaluit, Baffin Island, in the Canadian Arctic; Longyearbyen, Svalbard in Norway; and Petropavlovsk-Kamchatsky, Kamchatka in the Russian Far East, now serve the tourism market with regularly scheduled flights. Backcountry adventures, such as mountaineering, rafting, kayaking, angling and hunting, are served by large fleets of fixed wing and helicopter charter services. One of the most popular Arctic tourist venues is flight seeing – viewing vast expanses of wilderness from the sky. All these activities are aggressively promoted as either shore excursion programmes or destination attractions by the cruise ship industry.

Further integration of flight schedules will better connect Arctic nations with each other and the rest of the world; airports are expanding to accommodate larger aircraft and more passengers, and the increased use of fuel-efficient jets offset rising energy costs. All these advancements will facilitate Arctic mass tourism and it is probable that the cruise industry will be a major beneficiary. Specific examples of the commercial air transport developments now being implemented include:

- In May 2007, the Norwegian airline company, Wideroe, extended their flights between Tromsø and Kirkenes to Murmansk, posing serious competition to Aeroflot Nord already operating the Murmansk-Tromsø route (*Barents Observer*, 1/16/07).
- In 2007, Russia's Transportation Ministry began developing a new airport infrastructure strategy for the whole country, in direct response to a rapid increase in air traffic. According to the Federal Air Transport Agency, in 2006 Russian airlines transported 38 million passengers on domestic and international routes, 8.5% more than in the previous year. Improvements in air transport infrastructure will directly affect access to tourist attractions.
- Denmark and Iceland recently expanded service between their two nations. Iceland Express now offers direct flights between Copenhagen and Akureyri, northeast Iceland, in summer and flies to 13 destinations from Keflavík international airport (*Iceland Review*, 6/4/07).
- Greenland is expanding the Kangerlussuaq terminal, widening the building and improving the flow of domestic and international traffic. The new terminal will also provide better conditions for security and luggage handling (SIKU News, 12/11/07).
- In April 2008, Finnair and Russia's largest airline, Aeroflot, began flights between Helsinki-Vantaa and Sheremetyevo airport in Moscow four times daily or 28 times weekly, and the number of

flights operated jointly by Finnair and Rossiya between Helsinki and St Petersburg will rise to 22 a week. In 2007, there were only two daily departures between Helsinki and Moscow. Both carriers use Airbus A320 series planes (NewsRoom Finland, 2/22/08).

- In 2008, the town council of Ísafjördur, the capital of the West Fjords, Iceland, agreed to challenge the Ministry of Transport to make improvements to the local airport so that it can be used as an international airport (Iceland Review, 1/12/08).

Improved access to the Arctic has consistently resulted in both increasing numbers and wider geographic distribution of tourists. As Arctic and sub-Arctic airports grow and become more efficient, all segments of the tourism industry increase their potential for growth.

Fishing and hunting

Seasonal concentrations in the Arctic of fish and trophy-size wildlife, and the relatively high probability of meeting them in attractive marine and wilderness settings, are enormously appealing to anglers and hunters. The Arctic's long history of rather primitive sport angling and hunting, mainly for the wealthy, changed radically in the mid-20th century. After WWII, the availability of reliable bush planes, four-wheel drive vehicles, fiberglass boats and efficient boat engines, combined with new personal wealth, popularised this market in the Arctic as elsewhere. Today, highly regulated sport fishing and hunting thrive throughout the region. In coastal communities, charter boat captains familiar with local waters guide sport anglers attracted to salmon, halibut and sea trout. Inland anglers and hunters are transported to camp sites by charter boat or bush plane. As commercial fishing regulations in Arctic waters become more stringent, or commercial fishing is banned altogether, local fishers are using their boats for the sport angling market, providing a living and sustaining a way of life that might otherwise be lost.

Local Arctic and sub-Arctic communities provide guides, provisions, accommodations, transport services and diverse hospitality services to a sport clientele that consistently pay premium prices and demonstrate strong loyalty to return for their seasonal pursuits. Arctic governments glean revenues from licenses, permits and other taxes, and these funds are, in varying proportions, applied directly to fishery and wildlife management. The fact that most of the income derived from these sporting activities remains in the community has garnered very strong local support for this type of tourism (Snyder & Stonehouse, 2007: Chapter 1). Examples include:

- According to the most recent *National Survey of Fishing Hunting and Wildlife Associated Recreation*, in 2006 the State of Alaska's sport fishing market hosted 172,000 non-resident anglers who spent a total of 838,000 days fishing. Their economic contributions to Alaskan communities totalled nearly $564 million and more than $413 million of this amount was spent on trip-related expenses such as food and lodging, transport and guide services, licensing and other trip-related expenses (US Fish and Wildlife Service, 2007).

- Iceland's citizens derive extraordinary revenues from its sport fishing market. A fishing license costs 200,000 Ikr per day, a sum that does not include a guide, transportation, or equipment, food, lodging and transport. For those essentials, Iceland's Tourist Bureau suggests a daily budget of US$2,000.
- In Scandinavia, trophy fishing has been a tradition for centuries. The large size of this industry is exemplified by Norway, where 518 tons of total river-caught fish were landed in 2005. But, unfortunately, the most recent fishery statistics reveal both the commercial and sport fishing catch for most species in Norway are in decline (Statistics Norway, 2008).
- Canada's Department of Fisheries and Oceans *2000 Survey of Recreational Fishing in Canada*, reports that 'Recreational fishing is an important economic activity in the natural resources sector. In total, anglers spent $6.7 billion in Canada in 2000. Of this amount, $4.7 billion was directly associated with recreational fishing. Anglers spent over $2.4 billion on trip expenses such as package deals, accommodation, food, transportation, fishing supplies and other services directly related to their angling activities' (Canada Department of Fisheries and Oceans, 2005).
- Each summer, 500 non-resident anglers are allowed to fish for king salmon and rainbow trout along the rivers of the Russian Federation's Kamchatka Peninsula at a weekly cost of between $2000 and $5000 per person, plus transportation. The United Nations Development Programme and the Wild Salmon Center are seeking to promote catch-and-release fishing as an integral part of multi-million dollar ecotourism development and habitat conservation projects. The Kola Peninsula is now actively promoting sport fishing throughout this region of the Russian Federation.

As mentioned previously, native people have hosted sportsmen in their communities for nearly two centuries. The economic and natural resource implications of that long-standing relationship have proven to be extremely beneficial to native people. Most importantly, the majority of sportsmen expenditures remain in the community, directly supporting native jobs and businesses. Wildlife and fishery management is either directly controlled or closely monitored by native people and thus benefit from this competent attention. As native people attain greater self-governance, the benefits of their unique relationship with sportsmen will be, most probably, a significant contributor to their economic self-sufficiency.

Nature tourism

Nature tourism caters for tourists observing wildlife in natural habitats, and experiencing the beauty and solitude of natural areas (Figure 2.5). Primary locations are the Arctic's many national parks, wildlife refuges, marine sanctuaries and World Heritage Sites. These include North America's largest protected area, embracing Canada's Kluane National Park, the USA's Glacier Bay and Wrangel St. Elias National Parks, and the Alsek Tatshenshini Wild River that flows through both countries. Northeast Greenland National Park is the world's largest single national park; Europe's largest park is Vatnajökull Park, established in Iceland in 2008; and Europe's largest international protected area is a jointly managed region including Finland's Oulanka National Park and Russia's Paanajarvi National Park. In addition to those massive land areas, much of the Arctic marine environment remains uncharted and a virtual wilderness of astounding proportion. For a list of polar national parks and heritage sites see Appendix B.

Expedition cruise ships, chartered wildlife-sightseeing boats, river rafting and sea kayaking cater for marine-based tourists. Land-based nature tourism involves guided backcountry trips through wilderness

Figure 2.5 MV *Endeavour*. The ship has nosed in close to the beach: passengers are transferring to Zodiacs for scenic cruising and a landing. Photo: Esther Bertram.

regions, or viewing from the windows of railroad cars and all-terrain vehicles. The main attractions are the beauty and solace of immense wilderness areas, and seasonally abundant wildlife, including massive migrations of birds and mammals that can be predicted with near-certainty. Birding, wildlife and nature photography, backpacking and wilderness trekking, kayaking and river rafting are some of the most popular activities. Seasons do not limit the attractions: cross-country skiing, snow shoeing and dog sledding are some of the recreational activities avidly pursued in winter.

Increased publicity regarding Arctic climate change and the environmental changes it is creating, both real and perceived, has accelerated the growth of the nature tourism market. For example, growing numbers of wildlife enthusiasts travel to Churchill, Canada, and Svalbard to view increasingly endangered polar bears and seals; birders are attracted to larger concentrations of species; and tourists are increasingly eager to see larger populations of whales, seals and sea lions. Simultaneously, the recession of ice is allowing greater access to backcountry regions and navigable rivers, while simultaneously expanding the length of the nature tourist season.

The economic benefits of the nature tourism market are enormous and its popularity is growing substantially. Comparison of the 2001 and 2006 wildlife-watching data for Alaska evidences this growth. According to the *National Survey of Fishing Hunting and Wildlife – Associated Recreation*, in 2001 Alaska hosted 420,000 wildlife watchers who spent nearly $499 million. By 2006, those numbers increased to 514,000 wildlife watchers who spent more than $705 million (United States Fish and Wildlife Service, 2002, 2007). Specialised guided wildlife tours are growing in popularity despite the relatively high costs, which range from $4000 to $8000 per person for a six-day holiday (Natural Habitat Adventures, 2008).

Adventure tourism

Adventure tourism provides a sense of personal achievement and exhilaration from meeting challenges and perils. Land-based adventure tourism includes mountaineering, hiking enormous distances, cross-country skiing along difficult routes and kayaking wilderness rivers with intense white-water challenges. The marine environment hosts adventurers exploring the coasts or attempting the Northwest Passage in all manner of watercraft, and sea kayakers attempting to circumnavigate Greenland.

Unlike guided wildlife tours, sport fishing and hunting, the adventure tourist is rarely required to obtain a backcountry permit, have a license to demonstrate competence or report their intended routes and schedules. In a few instances, reaching the summit of popular Arctic peaks or

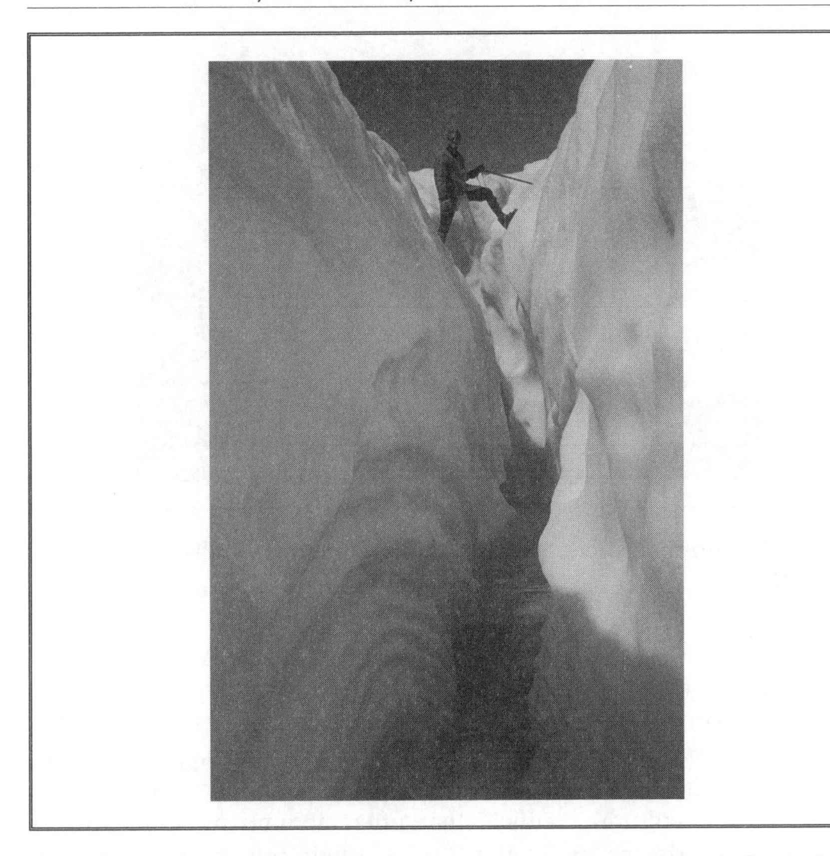

Figure 2.6 Tourists and emergency response agencies are challenged by adventure tourism. Photo: JMS.

traversing ice fields has resulted in the need for permits to reduce congestion, but these are the exceptions. As a consequence, Arctic adventure tourism is almost impossible to manage. Harsh environmental features and weather conditions combined with unreliable communication and remote emergency services can be fatal. But the adventure tourism market is undeterred by those hazards. Its growth is fueled by specialised publications, organisations, and personal networks vigorously supplying the adventurers with new challenges, tales of recent achievements and the availability of new technologies.

Culture and heritage tourism

Culture and heritage tourism caters for visitors who want to experience local history, art and cultural traditions. This topic is dealt with more fully in Chapter 7.

Figure 2.7 Tourism is strengthening native culture by revitalising native art. Photo: JMS.

According to a 2003 study by the Travel Industry Association and *Smithsonian Magazine*, 118 million persons worldwide sought history and culture tourism experiences, an increase of 13% from 1996 (Olson, 2003). Arctic native people are actively pursuing culture heritage tourism to strengthen their economies and cultures, while simultaneously attempting to reduce cultural conflicts that often accompany this form of tourism. The primary motivation for pursuing cultural tourism is to achieve the economic self-sufficiency needed to sustain recently acquired political autonomy. The Home Rule Government in Greenland, the Alaska Native Corporations, Nunavut self-rule in Canada, expanded Saami governance of their resources, and new self-governance opportunities for Russia's indigenous people exemplify the new political and economic realities of the Arctic. The creation of cultural tourist attractions and the jobs, income and sales revenue they generate are positive contributions, while inappropriate visitor behaviour, technological innovations and the intrusion of large numbers of people seriously challenge the preservation of cultural integrity.

Alaska Native Corporations own and operate charter boats that visit ancestral settlements and provide wildlife-viewing opportunities. In the Canadian Arctic, native people's marine tourism ranges from kayak journeys to cruise ship operations. Most recently, Inuit-owned Cruise North Expeditions began service in summer 2008, with Arctic trips that highlight wildlife, including polar bears, caribou, bearded seals, bowhead and beluga whales, musk ox and walrus. Its itineraries include a new route to Ellesmere Island, the most northerly island in the Canadian Arctic archipelago (*Canadian Press*, 11/20/07). Inuit legislators of Greenland's Homeland Government are implementing an ambitious tourism economic development strategy that is heavily dependent on the growth and expansion of cruise ship operations. The Saami share their reindeer herding and other customs with tourists. All the Arctic cultural venues seek additional revenues through the sale of art, crafts, goods and services.

The enormous geographic scope of these markets deserves emphasis. All are hosted by the eight Arctic nations, in geographical areas including ice-infested seas that exhibit some of the world's most severe maritime conditions, and lands that are among the world's largest expanses of true wilderness. While these environments are clearly attractive to tourists, their lack of human infrastructure severely constrains both resource management and human emergency response capabilities. Visitor safety and behaviour, the ways in which natural and cultural resources are utilised, the seasons and duration of resource use, the geographic distribution of tourism activities and their inherent dangers, vary from one segment of the Arctic tourism market to another, but all present substantial problems for tour operators and governments alike.

Summary and Conclusions

Though unlikely venues for tourism, both polar regions have histories of recreational visits going back almost two centuries for the Arctic and over half a century for the Antarctic. Both currently support tourist industries that are growing and diversifying rapidly.

Early Arctic tourists were 19th-century wealthy yachtsmen and hunters, soon to be followed by back-packing travellers attracted by the easily accessible wide open spaces and wildlife. Mass tourism to the Arctic began later in the 19th century with the advent of railroads and shipping services, expanding further during the early 20th century with the advent of cruise touring, and further again after World War II with the development of mass air transport.

Currently, Arctic tourism is dominated by scenic cruising, but with substantial sectors of airborne travel, sport fishing and hunting, and nature, adventure and cultural and historic tourism. The industry is controlled by the governments of eight nations, all welcoming and encouraging it by providing infrastructure and financial support. Tourism provides much-needed revenues for indigenous communities, more stable and relevant than jobs in extractive and other industries, and to some degree supportive of their cultures. Native people welcome it so long as it is not thrust upon them, and they can maintain a measure of control.

Chapter 3

Antarctic Tourism: History and Development

Introduction: A Late Starter

Antarctic tourism began much later than Arctic tourism, and followed a completely different pattern of development. This is hardly surprising: long after the Arctic became known to developers, including tour operators, Antarctica itself remained a mystery – a continent believed to exist at the bottom of the world, which was gradually outlined by research expeditions during the early to mid-19th century. When examined, it proved far less accessible or user-friendly than the Arctic – more remote from northern hemisphere civilisation, more difficult to reach and lacking such amenities for travellers as trappers' huts, local outfitters and friendly natives who were willing to guide. The unfolding Antarctic was clearly no place for hikers, hunters and lovers of recreational solitude of the kind that had pioneered Arctic tourism.

As late as 1895, when commercial tourism was already well established in the Arctic, the Sixth International Geographical Congress in London recorded its opinion that:

> the exploration of the Antarctic Regions is the greatest piece of geographical exploration still to be undertaken. That in view of the additions to knowledge in almost every branch of science which would result from such a scientific exploration the Congress recommends that the scientific societies throughout the world should urge... that this work should be undertaken before the close of the century. (Quoted in Mill, 1905: 384–385)

Turn-of-the-century expeditions from Belgium, Britain, France, Germany and Sweden took up the challenge, and Scotland sent an expedition of its own. Their reports and popular accounts for the first time in many years brought Antarctica into public awareness. However, they told of discomforts and hardships that were unlikely to encourage visits for pleasure or recreation. A further half-century would pass before Antarctic tourism began.

For a tourist guide to Antarctica and the southern islands see Rubin (2000). For a guide to the South American sector of Antarctica, see Stonehouse (2006). Soper (1994) provides an illustrated guide to Antarctic wildlife. Bertram (2007) reviews an expanding industry, and Bertram and

Stonehouse (2007) review its current rôle in tourism, which is also discussed in Chapter 8.

Antarctic Tourism: The Beginnings

Codling (1995: 167–177) and Stonehouse (2007: 1004–1007) have surveyed the early history of Antarctic tourism. Through much of the early to mid-19th century, travel from northern ports to the southern hemisphere was functional rather than recreational. Sailing-ship passages to the far east and antipodes involved spartan conditions on long voyages through the Roaring Forties, almost inevitably in rough seas and sometimes travelling far enough south to encounter icebergs. From 1879, Thomas Cook and Sons began to advertise tours to Australia and New Zealand that stressed travel for pleasure, in fast steamships designed for the comfort of cruise passengers.

In 1910, responding to popular interest in the Antarctic expeditions, the same company advertised a cruise from New Zealand to the Ross Sea sector of Antarctica, where Robert Scott and Ernest Shackleton, the most prominent British explorers, had recently based their expeditions. A Christchurch (New Zealand) newspaper noted:

> There is a possibility of the Antarctic regions being visited by a party of tourists next year, Messrs Thomas Cook and Sons having put forward proposals for the despatch of a vessel to McMurdo Sound. The trip, it is estimated, will take fifty days, and it is intended that the vessel should leave some New Zealand port about the end of the year, so as to arrive at the Antarctic in mid-summer. (*The Press*, 4/11/1910)

Although 'some members of the New Zealand Parliament, a number of ladies, and several gentlemen' enquired about the proposal, there is no record that such a pioneering cruise actually took place. Later proposals during the 1920s and early 1930s advertised guidance by veterans of the expeditions, but were no more successful (Snyder, 2007: 27–28).

During these decades, tourism existed south of the normal shipping routes, but was restricted to small numbers of passengers in ships that were making annual servicing voyages to such remote southern outposts as Macquarie Island, the sub-Antarctic islands south of New Zealand, and the solitary Argentine meteorological station, Orcadas, on the South Orkney Islands – at that time the only permanently inhabited human outpost in the Antarctic region. From 1904, ships of the whaling industry arrived, setting up operations mainly from stations on South Georgia and the South Shetland Islands, and from factory ships further south. Transports again from time to time carried passengers as interested observers. The Norwegian whale factory ship, *Thorshavn*, carried the captain's wife, Karoline Mikkelsen, as a passenger, who by landing on the Ingrid Christensen Coast of Dronning Maud Land, in February 1935, became the first women known to have stepped onto the Antarctic continent (Bogen, 1957, quoted in Headland, 1989: 294).

SS *Fleurus*, a converted trawler owned by a Norwegian whaling company, was chartered by the Falkland Islands government to carry mail to whaling stations in the Falkland Islands Dependencies. With two staterooms amidships, *Fleurus* sold tickets to the public for return voyages between the Falkland Islands and the Dependencies, and might reasonably be considered the first Antarctic cruise ship (Hart, 2006: 221). In 1927 and 1928, it carried the governor of the Falkland Islands on fact-finding tours of the southern whaling grounds, including South Georgia, the South Shetland Islands and Graham Land.

The Antarctic tourism industry we know today dates from the mid-1950s. According to Snyder, it originated in:

> the unique combination of favourable world-wide publicity associated with the International Geophysical Year (IGY), the availability of modern transport technologies, a pent-up demand for recreation, and new personal wealth. The announcement and implementation of the IGY and the first commercial jet and cruise-ship transport of Antarctic tourists all occurred between 1956 and 1958. (Snyder, 2007: 29)

Potential travellers and tour operators quickly realised that safe, commercially viable visits to the Antarctic region would be possible only for large organised parties in dedicated ships or aircraft.

Early flights from 1956 and cruises from 1958 are recorded in Headland (1989) and tabled in Reich (1980: 21, 207–208) and Enzenbacher (1992: 18). The first recorded tourist flight over Antarctica was made by a Lan-Chile DC6-B operating from Tierra del Fuego, which overflew the South Shetland Islands and Antarctic Peninsula with a group of passengers. In 1957, a Pan-American Stratocruiser chartered by the US government took 160 US naval personnel and a number of civilian passengers from Christchurch, New Zealand, to land at McMurdo Station, McMurdo Sound.

The first tourist cruise ship of this period (for the first-ever, see *Fleurus*, above), the Argentine naval transport, *Les Eclaireurs*, in 1958 carried 194 passengers in two voyages from Ushuaia (Tierra del Fuego) to Antarctic Peninsula and the Scotia Arc. In the following year, two similar voyages by Chilean transports, *Navarino* and *Yapeyú*, took a total of 344 passengers to the same area.

None of these early ventures proved popular or profitable enough to start regular services: no further tourist cruises to Antarctica were recorded until 1966, and no further flights until 1968. Cruises had depended mainly on local demand in South America, which may have proved limited. Overflights to the South American sector depended for their success on good visibility over Bransfield Strait, which, then as now, could seldom be guaranteed, and there were no safe all-weather landing facilities for large aircraft available in the area. Flights to the Ross Sea sector required the goodwill of the US government, which did not see any reason to encourage tourism. Landing facilities in McMurdo Sound were not made generally available to commercial operators.

Early shipborne tourism

Regular cruising to Antarctica began with the advent of the US travel operator, Lindblad Travel Inc., under the personal direction of a Swedish-American entrepreneur, Lars-Eric Lindblad, who specialised in small-group travel to out-of-the-way places. In early 1966 and again in 1967, Lindblad chartered the Argentine naval transport, *Lapataia*, offering adventure cruises in the Peninsula sector to his mainly North American clientele. In 1968, he chartered the Chilean passenger liner, *Navarino*, for two similar cruises, and the Danish ice-working ship, *Magga Dan*, for two cruises from New Zealand to McMurdo Sound. In 1969, he chartered another Chilean naval transport, *Aquiles*, for a further cruise to the Peninsula.

As on all his cruises to remote corners of the world, Lindblad made a point of visiting harbours and beaches where passengers could land, using either the ships' tenders, or preferably flat-bottomed inflatable boats with outboard motors (generically termed 'Zodiacs') to take them ashore. Thus, his Antarctic cruises, usually from southern South America and lasting eight to twelve days, were characterised by landings in which passengers came into close contact with basking seals, colonies of penguins and other wildlife, often on shores that had no previous record of human visits.

The 'Lindblad pattern' (Stonehouse & Crosbie, 1995: 221–222) operates best with groups of up to 140 passengers, guided ashore and afloat by staff with Antarctic field experience. Each voyage becomes an 'expedition' with lectures, briefings and shore landings. Passengers are particularly briefed on behaviour ashore, including possible dangers, and the need to avoid interference with flora, fauna and scientific research projects. Issued with bright-red padded jackets to ensure that (a) they are adequately dressed and (b) can be seen from a distance, they are landed in parties of 10–14, up to a total of 100 ashore at a time. They are mostly free to wander within a prescribed area (some operators delimit the area with traffic cones), but are required to stay within sight and easy reach of the landing point. Zodiacs are also used for scenic cruising among icebergs and along the shores. Evenings are usually occupied by 'recap' sessions to discuss the day's activities. Stonehouse and Crosbie comment:

> This pattern of shipboard education, linked with guidelines that require visitors to avoid walking on vegetation, disturbing nesting birds and leaving litter, has recommended itself strongly to the kinds of tourists who have so far made up the majority in Antarctica. Many claim that they would avoid tours that did not feature similar levels of concern. As a consequence of the Lindblad pattern, in an environment that has proved vulnerable to impacts from scientists, and that many regard as hypersensitive to visitor impact of any kind, there is so far remarkably little evidence of damage from tourism. (Stonehouse & Crosbie, 1995: 222)

That comment, written almost 30 years after Lindblad's first voyage, remains equally true today. His pattern of management set objectives that ensured high standards of behaviour among tour operators and tourists alike, resulting in far less environmental damage than might have been expected had Antarctic tourism developed without it.

Travel-writers and journalists alerted scientists and conservationists to possibilities of serious disturbance to the region's 'pristine' environment, and to the scientific research that, under the recently inaugurated Antarctic Treaty, was designated the most important single objective of human endeavour in Antarctica. They did not, in general, distinguish between Lindblad's small-ship operations and a number of Argentine, Chilean and Spanish cruises that continued to visit the Peninsula area from 1968 into the mid-1970s, involving ships carrying up to 400 passengers, with a less-disciplined approach to landings. The last of these, the Italian cruise liner, *Enrico C*, with accommodation for 700 passengers, made a single voyage to the Peninsula in 1976–1977.

Thereafter, the market for big-ship cruises appears to have vanished, though only for the time being. Meanwhile, Lindblad continued. Finding that chartered ships imposed serious limitations on the way he preferred to work, he commissioned the construction of a dedicated cruise ship of 2500 tons with double-hull, which would operate safely in polar waters as well as worldwide. *Lindblad Explorer* (Figure 2.4) came into service in 1969–1970, accommodating its owner's ideal complement of 90–100 passengers, with a crew and staff of about 60 (Lindblad & Fuller, 1983: 105, 151–154). For illustrated accounts of *Lindblad Explorer* voyages in Antarctic and other waters, see Snyder and Shackleton (1986).

During Antarctic tourism's first quarter century there was no central authority to record numbers of ships, voyages made and passengers carried each year, and basic data have already been lost. Figures provided by Reich (1980: 207–208) and Enzenbacher (1992: 18), though differing slightly in detail, give the best available estimates. For most of these years, passenger numbers are uncertain. In lieu, both authors indicate numbers of berths available in the ships, which provide upper limits but may not represent the numbers actually carried. Estimated numbers of cruises and passengers during this early period are summarised in Table 3.1.

As indicated in Table 3.1, between 1957 and 1980, Antarctic shipborne tourism:

- began with two seasons in which naval transports visited the Peninsula area, followed by a break for the following six seasons;
- re-started in 1965–1966 with the appearance of Lindblad Travel Inc., which operated almost alone during the three seasons to 1967–1968, with smaller ships carrying up to 60 passengers;
- involved between 1968–1969 and 1976–1977, small ships operated mainly by Lindblad, and larger ones, mainly under South American management, of up to 800 passenger capacity;
- reached peak numbers of 11 voyages and over 4000 total passenger capacity in 1974–1975;
- from 1976–1977 to 1980 was represented only by two ships, respectively, of 200 and 92 passenger capacity;

Figure 3.1 Sound Antarctic tourism: A well-behaved visitor to South Georgia watches elephant seals, and is in turn watched by newly-moulted king penguins. Photo: BS.

Table 3.1 Numbers of ships, voyages and passengers carried or berths available on Antarctic cruise ships, 1957–1958 to 1979–1980

Season starting	*Ships*	*Voyages*	*Passenger numbers or berths*
1957–1958	1	2	194
1958–1959	2	2	344
1965–1966	1	1	58
1966–1967	1	2	94
1967–1968	3	5	257
1968–1969	2	5	1,712
1969–1970	2	4	972
1970–1971	2	4	943
1971–1972	4	6	984
1972–1973	3	8	2,074
1973–1974	3	5	1,876
1974–1975	3	11	4,012
1975–1976	2	10	2,250
1976–1977	3	6	1,068
1977–1978	3	9	845
1978–1979	2	7	1,048
1979–1980	2	6	855
Total	–	93	19,586

Source: Based on Reich (1980) and Enzenbacher (1992). See also Figure 3.3

- involved a total of 93 voyages, of which 86 were to the Peninsula area, only seven visiting the Ross Sea sector.

To this record must be added an unknown number of visits, no doubt accompanied by landings, by yachts and other small craft operating mainly from South America. Some were privately owned and operated; others carried small numbers of paying passengers. In the absence of Antarctic ports or port authorities, there are no reliable records of such voyages. There are unlikely to have been more than two or three per year, bringing one or two dozen passengers to swell the statistics. More recently, small-craft numbers appear to have increased, but are still negligible in comparison with the number of registered cruise ships and their passengers.

Early airborne tourism

Following an 11-year break in tourist flights to Antarctica, a single flight – a one-off non-commercial venture organised by an academic institution, the Richard E. Byrd Polar Center in Boston, MA – was permitted in 1968 to use US government facilities in McMurdo Sound. Seventy-five passengers landed from a Convair 990 at Williams Field, McMurdo Sound, then continued over the South Pole to Rio Gallegos, Argentina, a distance of 8325 km (Reich, 1980: 209). Reich's table (1980: 211) records six further flights involving small numbers of tourists or adventurers in 1970–1974, followed by a three-year break.

In early 1977, a series of tourist overflights began from Australia and New Zealand to East Antarctica. These proved extremely popular; according to Bauer:

> The first series of sightseeing flights from Australia took place on 13 February and 16 March 1977, and flew over Macquarie Island and the South Magnetic Pole as well as parts of Victoria Land, Oates Land and George V Land. Quantas aircraft were used for the flights. They were organized by former *Australian Geographic* publisher Dick Smith. (Bauer, 2007: 190)

Smith, an enterprising businessman with strong geographical interests, sought originally to charter one such flight from Quantas. The resulting publicity stimulated general interest and started the series (Reich, 1980: 210). Flights from Australia used Boeing 747 aircraft and overflew the coast immediately south of Australia, mainly between Terre Adélie and Cape Adare. Those from New Zealand used DC10s and flew along the Victoria Land coast into McMurdo Sound, where they overflew Mount Erebus and US and New Zealand scientific stations. Bauer continues:

> The series of 30 Quantas and ten Air New Zealand flights that had carried some 10,000 passenger concluded on 16 February 1980, when the last Quantas plane returned from Antarctica. The end of the first series of overflights came as the result of the crash of Air New Zealand flight TE 901 in Antarctica on 28 November 1979, (commonly referred to as the Mount Erebus disaster) that claimed the lives of all 237 passengers and 20 crew. (Bauer, 2007: 190)

For an account of the Mount Erebus disaster and its aftermath in the New Zealand courts, see Mahon (1984).

Thus, during the first quarter-century of Antarctic tourism, airborne tourism was restricted to 43 overflights, which by Reich's (1980: 213) calculation brought 11,045 visitors to view the continent from the air. There were, as yet, no indications of the later and ecologically more significant development – landing passengers in remote areas of the continent for trekking, climbing and adventure (pp. 57–58). Overflights from New Zealand were not resumed after the disaster: those from Australia began again in 1994. They have continued from South America and have since been initiated from South Africa (p. 57).

Antarctic Tourism Today

In all but the first years of its existence, Antarctic tourism developed within the judicial framework of the international Antarctic Treaty of 1959, which was ratified and came into force in 1961. During the early years of the Treaty, successive two-yearly Antarctic Treaty Consultative Meetings (ATCMs) paid little direct attention to the industry growing within their newly claimed jurisdiction. Thus, for over three decades, Antarctic tourism developed almost without government-imposed constraints, and with few restrictions other than those imposed by the environment and the industry itself.

Fortunately, the industry was never able to diversify radically or fragment into dozens of small-scale operations, in the manner of Arctic tourism. This would almost certainly have raised serious environmental issues and safety problems. Though still growing, it has remained in the hands of a small number of operators, and comparatively restricted in its range of activities. More fortunately still, in 1991 there arose a responsible trade association within the industry, the International Association of Antarctica Tour Operators (IAATO). Initiated and guided by Lars-Eric Lindblad, IAATO was intended from its inception to 'act as a single voice in concerns of tourism and to advocate, promote and practise environmentally responsible private-sector travel to Antarctica' (Landau & Splettstoesser, 2007: 198).

From the start, IAATO showed responsibility toward the environment in which its members were working by developing guidelines both for tour operators and for their client visitors. The first-ever published code of conduct for Antarctic travellers, originating from four experienced field guides (Naveen *et al.*, 1989: see also Stonehouse, 1990) were developed by IAATO into practical guidelines – guidelines that were later adopted in principle by the treaty parties when they began to assume control (Appendix D). These issues are discussed further in Chapter 8.

Despite its freedom from judicial constraints during the early days, Antarctic tourism began simply and developed linearly, restricted mainly by the environment in which it operated. Where tourism in the Arctic has proliferated into many forms (outlined in Chapter 2), Antarctic tourism remains almost entirely confined to four basic sectors:

- shipborne tourism in small or medium-sized cruise ships that land their passengers for visits along the shore;
- shipborne tourism in large cruise ships that make no landings in the region;
- overflights from neighbouring continents that do not land in Antarctica;
- flights that land parties in Antarctica for brief visits, or for climbing, trekking and other activities.

To these may be added a fifth category – adventure tourism – that requires either ships or aircraft to bring it to Antarctica, but then develops in its own directions outside the classic patterns. As such, it represents a further challenge to any authority that may claim to have tourism in Antarctica under control.

Shipborne tourism with landings

While the first few years of Antarctic cruising involved large passenger liners, Lindblad's first small-ship visit in 1966 marked the start of a new mode. Small ships carrying 80–200 passengers, making several landings during the course of the voyage and in other ways practising the 'Lindblad pattern' of management, dominated the industry for over a quarter of a century. However, recognising the limitations of scale imposed by small ships, in December 1990, Lindblad introduced the larger French-operated cruise liner, *Ocean Princess*, with cruises involving a nominal 250 passengers drawn from a less adventurous and less physically active clientele.

Although *Ocean Princess* had capacity for over 700, Lindblad limited numbers, continuing his emphasis on information and education, but retaining the evening shows and other entertainments normally encountered on big-ship cruises. Zodiacs offered a limited number of landings with passenger numbers restricted to 100 at a time – a restriction imposed primarily by the need to return shore parties aboard quickly should the weather change. This pattern proved profitable enough for other ships of similar size to emulate during the 1990s, and has continued into the present century, providing a popular alternative to small-ship cruising.

Though cruises involving still-larger ships without landings are now available (see below), cruising by ships of 50–500 passenger capacity, with landings at selected sites, using fleets of inflatable boats with outboard engines, remains by far the largest sector of Antarctic tourism. Bertram (2007: 152–153) regards landings as arguably the most invasive of all tourist activities. However, all the ships within this range continue to operate the Lindblad pattern fairly closely. Those carrying 300–500 passengers necessarily take much longer than the smaller ships to accomplish a landing, thereby making fewer landings in the course of a voyage. However, all passengers are briefed beforehand on hazards and standards of behaviour, and accompanied ashore by guides who provide information and check safety.

Figure 3.2 Inflatable boats with outboard engines ferry a dozen passengers at a time from cruise ship to shore. Waterproof pants and warm anoraks are essential. Photo: BS.

At the smaller end of this market are yachts and motor vessels carrying up to 20 passengers, often catering for special-interest groups of photographers, climbers or campers. Numbers are uncertain, because only those that operate commercially (i.e. for charter or advertised cruises) qualify for IAATO membership and inclusion in their statistics. Others may be traced through gateway port records, but not all bother to register their intentions or movements precisely enough to be included. They are unlikely to add up to more than a few dozen ships per year.

For numbers of ships, voyages and passengers involved in tourism from 1992 to 1993, when IAATO records began, see Landau and Splettstoesser (2007: 201). Figure 3.3 illustrates the overall growth of passenger berths and numbers from the earliest voyages to the present.

The underlying Lindblad pattern of tourist management (p. 48) afloat and ashore has barely changed since the 1960s. Innovations have been introduced by operators catering for special-interest groups, including overnight camping, climbing, glacier walking, marathon running, canoeing among pack ice – all designed to attract younger and more active

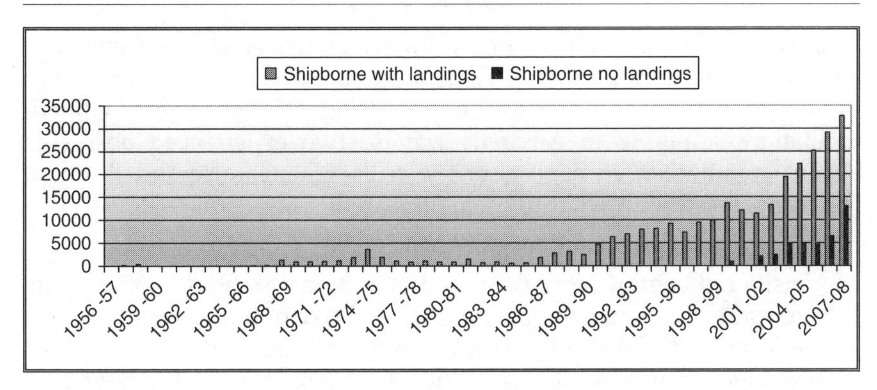

Figure 3.3 Estimated numbers of passengers on Antarctic cruise ships, 1956–2008. In 2008–9 numbers of landing passengers fell to 27,206, non-landing to 10,6520
Source: Bertram 2007; IAATO.

participants than those for whom the pattern originally catered. Environmental implications of this diversification are discussed in Chapter 8.

Shipborne tourism without landings

In 2000, the shipping line, Holland America Westline Tours Inc., introduced a third pattern of shipborne tourism – a liner of about 1000 passenger capacity engaged only in sight-seeing without landings. The arrival in subsequent years of more liners of the same size, plus still larger ones carrying 2500 or more passengers, has continued this trend (see below).

The appearance of MV *Rotterdam*, though fully sanctioned by the US Environmental Protection Agency as meeting all legal requirements under the US Antarctic Science, Tourism and Conservation Act of 1996, was regarded with misgivings by many for bringing a large ship with no ice strengthening to Antarctic waters, and by others for bringing an alien culture of bingo and showtime entertainment to the last unspoiled continent. In more realistic terms, it was merely an extension to Antarctic waters of traditional big-ship cruising, for long practised around South America. It catered for a clientele to whom a lively and entertaining shipboard life was a major attraction, and who welcomed the addition of opportunities to appreciate Antarctica for three to four days, from either the comfort of their staterooms or the superb viewing platform of open decks 30 m above the sea.

Though MV *Rotterdam* had not previously visited Antarctica, the operating company in its application pointed out that several officers aboard had gained ice navigation practice in scheduled cruises to Glacier

Bay, Alaska, and the ship carried a highly experienced ice-master. The company chose to follow the stringent operating regulations that apply in Glacier Bay, in addition to the less stringent, indeed less defined, regulations applying to Antarctic waters. Two experienced observers aboard commented on aspects of the cruise: Stonehouse and Brigham were impressed with what to them appeared:

> a safe, carefully planned, and well-executed operation, with well-considered contingency plans, and operational standards relating to environmental protection that far exceeded those required in conforming to generally accepted Antarctic guidelines. The cruise provided enjoyment and education for more than 1600 passengers and crew, and environmental impacts were minimal. (Stonehouse & Brigham, 2000: 348)

Antarctic cruising without landing quickly proved popular. From 2002, more companies and ships included Antarctica in their itineraries, and both the market and the size of ships involved continued to grow.

In 2007, the still-larger liner, MS *Golden Princess*, certified to carry up to 3100 passengers and a crew of 1060, made a first voyage to Antarctic Peninsula. Princess Cruises, a US-registered company, had previously operated three smaller liners carrying up to 1500 passengers in Antarctic waters. Aspects of the *Golden Princess* cruise were discussed by Bertram *et al.* (2008: 177–180), who considered it carefully planned, unhurried and well-executed, with all environmental protection requirements and constraints closely observed. As guest lecturers aboard, Bertram and Gunn were given ample facilities and high priorities to educate both passengers and crew, including lectures and presentations in the theatre (seating c. 800 and usually attended to capacity) and commentaries from the bridge during daylight hours.

> Though competing for attention with many other shipboard activities, from bingo to ballroom-dancing instruction, Antarctic presentations were given priority immediately before and during the four days in the far south. All passengers with whom the authors exchanged views seemed keen to learn about the Antarctic, and showed genuine interest and concern over the future of the continent. The impression was that the educational element played a special role which the company was concerned to foster, one that may readily be maintained, encouraged and developed on all such cruises. (Bertram *et al.*, 2008: 179)

Insofar as it attracts a different clientele (including older and less mobile passengers) from those who favour smaller-ship cruises with landings, and also provides a range of less expensive passages, this market seems likely to continue its expansion. Despite misgivings that big ships are potentially more hazardous than small ones, no convincing case against the use of large cruise liners has yet been made. Bertram *et al.* continue:

> for Antarctic cruises [operators] employ experienced ice pilots, follow conservative itineraries in well-chartered waters, allow adequate time to complete their passages at safe speeds, and carry all the equipment required under international conventions to deal with oil spills and other emergencies. (Bertram *et al.*, 2008: 179)

However, they point out that a large cruise liner foundering in Antarctic waters would have a potential for disaster unprecedented in the history of the region. They comment also that a recent Antarctic Treaty resolution raising questions concerning Antarctic shipborne tourism generally, and the presence of large ships in particular (United States, 2007), 'can only be welcomed by all for whom the seemingly unlimited growth and diversification of Antarctic shipborne tourism are matters for concern'.

Airborne tourism: Overflights

Overflights continue to prove attractive: for a recent review, see Bauer (2007). In a survey of all aspects of Antarctic tourism based mainly on IAATO records, Bertram (2007: 160) comments that numbers of passengers carried in overflights fluctuate widely from year to year, but show no overall tendency to increase. Flights from Punta Arenas, Tierra del Fuego, to the Antarctic Peninsula region, often planned as excursions for cruise ship passengers to the area, are frequently cancelled due to bad weather, but highly rewarding when they occur. On clear days, passengers have panoramic views of all they are likely to see during their cruise.

The establishment of an air link between Hobart, Tasmania and Casey, a permanent station in Australian Antarctic Territory (Powell & Jackson, 2007: 50–51) offers possibilities for further development of flights from Australia. Similar flights appear to be operated over Dronning Maud Land by Antarctic Logistics Company International (ALCI) and The Antarctic Company (TAC), both from South Africa.

Airborne tourism with landings

Tourist flights to Antarctica that involve landings were slow to develop commercially because of high overhead costs and lack of infrastructure and amenities for passengers in Antarctica. They increased gradually during the 1980s and 1990s, and since 2003 have expanded more rapidly. Bertram (2007) traces their origins to 1984, when the Chilean government began marketing short-stay trips to their bases on

King George Island, South Shetland Islands. Discovery of widespread 'blue-ice areas', i.e. windswept areas of inland ice (Mellor & Swithinbank, 1989) on which wheeled transport aircraft could operate, opened up much of the continent both to scientific investigation and to tourism.

From 1987, the private company, Adventure Network International (ANI), had already begun operating in the Patriot Hills area of West Antarctica, 'pioneering the concept of landing transport aircraft on naturally occurring snow-free glacier ice where ski aircraft are the norm' (Swithinbank, 1997: 243). Throughout the 1990s, ANI provided flights between Punta Arenas and Patriot Hills, Ellsworth Mountains (80°S, 81°W), mainly for mountaineers and adventurers. From Patriot Hills, clients were flown on to Vinson Massif, the South Pole and other attractions. Since 2003–2004, when ANI was bought by Antarctic Logistics & Expeditions (ALE), the combined ALE/ANI has continued to cater for adventure tourism:

> mainly in the continental interior. Clients include mountaineers, trekkers and visitors who want to experience Antarctica remote from the cruise-ships' beaten track. It also provides emergency rescue services for other tourist activities, and is available to support other private or national expeditions. (Bertram, 2007: 157)

Polar Logistics, a development of ANI, was established in 1991 mainly to carry personnel and cargo on government operations. In 1996, it pioneered direct flights by Hercules aircraft between Cape Town and Dronning Maud Land, primarily to provide services for German, Indian, Japanese, Norwegian, Russian, South African and Swedish stations, but simultaneously opening another possible route for airborne tourism into the heart of Antarctica.

A more recent development has been flights by a Chilean company from Punta Arenas to connect with cruise ships at the Chilean air station, Teniente Marsh, King George Island, South Shetland Islands. Taking little more than two hours, this essentially eliminates the two-day crossing from South America to Antarctic Peninsula by ship, via the Drake Passage, which many passengers find tedious in rough weather. From the South Shetlands onward they may sail more comfortably in sheltered and calmer waters.

Antarctic adventure tourism

Lamers *et al.* (2007) review this form of Antarctic tourism that has arisen in recent years, and appears to be growing. In attempting a definition, they make the general observation that adventure tourism 'entails an interaction between a participant and an environment in which the outcome is uncertain', an uncertainty that often translates into risks for participants. As scientific work and mainstream tourism in Antarctica seek to exclude risks wherever possible, a form of tourism that embodies risks runs contrary to the philosophy and practices of the continent's other users.

This matters because, in an environment where the need to help others in difficulties is generally recognised, risk-takers impose unnecessary constraints on the work or enjoyment of others. If sky-divers, ill-equipped mountaineers or adventurers who are seeking to cross the continent unaided get into difficulties, they have to be helped, perhaps at the expense of a more responsible scientific or tourist operation. Adventure tourism tends, therefore, to be regarded as self-indulgent and to be avoided in Antarctica.

Lamers *et al.* (2007: 172) distinguish three categories into which Antarctic adventure tourism currently falls:

- ship-based, operated from cruise ships along coasts;
- land-based, operated from tented camps serviced by aircraft;
- independent adventurers arriving by private boats or aircraft.

They point out that tour operators who are members of IAATO (p. 53), and offer adventure tourism as extensions of their more orthodox seaborne or air-based tours, have developed:

> guidelines and procedures for passengers, staff and crews... that address the activities' specific risks. Prior to any trip, tour operators screen the passengers for physical and mental competence, and for experience in the particular sports involved. In addition, participants are required to declare in writing that they accept the risks involved in the activity.

In addition, IAATO ensures that member companies are well insured, and that they are capable of dealing with incidents with minimal reliance on facilities and support from uninvolved national programmes. Most experienced in this field are the air tourism operators, ANI/ALE (p. 58), that have offered logistically well-planned and generally safe 'adventures', and support for independent groups of 'adventurers', in Antarctica for almost two decades.

Independent adventurers operating outside the IAATO network of advice and practical help are, theoretically at least, more likely to get into difficulties than experienced operators within. Their numbers are unknown, but almost certainly very few. Adventure tourism that works both within and outside the system is currently attracting concern among policy makers (Chapter 8). Overall, Lamers *et al.* conclude that, though at present a very small niche market, it is one that may be on the eve of rapid growth, and thus merits the full consideration currently being afforded:

> The level of attention that adventure tourism gets from policy makers may seem unjustified at first sight, but [is] warranted from a longer-term perspective, particularly in view of the slowness of Antarctic decision-making processes on tourism. (Lamers *et al.*, 2007: 183)

Antarctic ecotourism

Ecotourism is defined and discussed more fully in Chapter 5. Here, we need only point it out as a well-structured approach to wilderness management, widely accepted in the north as providing criteria and desiderata for sustainable tourism, and particularly appropriate for visitor management on the beaches of Antarctica. Indeed, the principles of ecotourism should be clearly apparent in any statement of both management regulations and tourist company practices. The degree to which these principles are met is discussed more fully in Chapters 8 and 9.

Summary and Conclusions

Because Antarctica remained undiscovered long after the Arctic was known and exploited, Antarctic tourism developed much later. Not until the mid-20th century were the first commercial flights and cruises available to passengers. Because there are no indigenous people, and the terrain is less friendly to visitors, Antarctic tourism has developed along narrow lines – tightly scheduled flights and cruises carrying large, carefully managed groups, with few opportunities for individuals or small adventure parties. We describe the industry under four main headings – shipborne tourism with and without landings, and airborne tourism with and without landings, to which is appended the minor fifth category of adventure tourism, which may be offered by standard operators or undertaken independently.

Antarctic tourism began in the late 1950s, in the years following the International Geophysical Year 1957–1958, with small numbers of over-flights and cruises from South America. Slow to develop during the early days, it began to grow especially during the 1990s, with shipborne cruising pre-eminent, overflights from South America, Australia and New Zealand, and much smaller components of land-based activities – the pattern that essentially operates today. Antarctica has no indigenous communities, and claims to sovereignty by seven nations are not generally recognised. Relations between the tourism industry and Antarctic Treaty System by which it is governed are discussed in Chapter 9.

Chapter 4

Tourism in Changing Polar Environments

Introduction: A Changing World

The longer we record natural phenomena, the more we become aware of environmental change. Ancestors with shorter life spans and little knowledge of the past might be excused for assuming that the world changed little from century to century and millennium to millennium. However, from the early 19th century, the geological record alone revealed vast worldwide changes through an immense span of time. More immediately, advancing and retreating glaciers, shifting coastlines, dynamic rivers, the clearance of forests for cultivation, and folk memories of harder or softer winters, warmer or colder summers, all indicated that radical changes could occur within human lifetimes. Climatic, archaeological and social historic records confirm that natural environments are constantly changing. Plant and animal populations adjust: humans have found that they must do the same.

This chapter discusses recent changes, both natural and man-made, which have affected polar regions within the last few decades, particularly those that affect polar tourism, and are thus part of the background to tourism management. For an early and wide-ranging assessment of possible consequences of global changes on the Arctic, in particular changes affecting Arctic societies, see summaries and individual abstracts in Lange and Flanders (1993). Here, we distinguish two kinds of changes:

- Those due primarily to cosmic events, including the constant redistribution of Earth's crust, and radical shifts in climate, some of which may be triggered or intensified by human activities.
- Those due primarily to human activities, whether for subsistence or commercial purposes, including hunting, prospecting, mining, landscape modification and tourism itself.

Changes due to Cosmic Events

Earth itself is dynamic: the surface movements that produced the familiar pattern of oceans and continents are a continuing process. Some, like the opening of mid-oceanic ridges and grating of crustal plates against each other, occur at rates measurable in millimetres per month. Others, like coastal erosion, occur rapidly enough to be noticeable within decades, certainly within human life spans. Others more rapid still, like earthquake shockwaves and volcanic eruptions, create sudden environmental catastrophes that alter whole regions in moments.

Crustal changes

Polar regions probably owe their frigidity to the movements of crustal plates over 60 million years ago (see below), and both currently have live tectonic areas along their fringes. In the north, these include Kamchatka, with 22 active volcanoes arising from the tundra, and the chains of mountains that include the Aleutian Islands and the southern ranges of Alaska. On the Atlantic side of the Arctic, just south of the Arctic Circle, Iceland is particularly active (Box 4.1): its hot springs and rumblings now contribute their share to Iceland's flourishing tourism.

Box 4.1 Iceland

Lying athwart the Mid-Atlantic Ridge, from which it has arisen during the past 20 million years, Iceland is the world's largest volcanic island (area 103,000 km^2), and one of its liveliest, with some 200 volcanoes scattered among its tundra meadows, concentrated in a SW–NE active zone that crosses the island. For details of its vulcanicity and economic consequences, see Münzer (1985). Stonehouse reports:

> The island is dotted with fissures and fumaroles which from time to time grumble and belch like angry walruses. Icelanders accept philosophically the appearance of new offshore islands, such as Surtsey, which arose with a roar off the south coast in 1963. They recalculate the country's total area, declare the new islands protected, and send ecologists out to study the flora and fauna that move in to colonize them. (Stonehouse, 1990: 167)

Just 10 years later, the nearby fishing port of Vestmannaeyjar was devastated by ash and lava from a neighbouring volcano. The townsfolk pumped seawater, cooling and directing the lava flows to protect and enhance the harbour, and within a year had rebuilt and reoccupied their town.

Continental Antarctica stands on a tectonic plate of its own, ringed by divergent plate boundaries along 95% of its perimeter (LeMasurier, 1990: 1). Though many peaks showing above the icecap provide evidence of past activity, current vulcanism is restricted to the Peninsula and Scotia Arc region in West Antarctica, and to the Ross Sea region in East Antarctica. East Antarctica's most celebrated volcano, Mount Erebus, with an almost constant plume of steam, is seen annually by thousands of shipborne and airborne tourists. More active are some of the islands of the Scotia Arc, off Antarctic Peninsula. Of 11 islands in the remote South Sandwich group, all but three have been reported active in recent decades. In the South Shetland Islands, Deception Island has a long history of activity, most recently in 1967–1970 when a series of eruptions raised columns of scoria and steam high into the air and destroyed two scientific stations. Yet the island is one of the area's main attractions for cruise ships, whose passengers are invited to bathe in hot springs along the beach.

Less spectacular changes in shorelines and coastal flats, particularly in polar areas, are due to tectonic uplift – local rising of land following the loss of massive ice sheets during the past few thousand years. Many Arctic and sub-Arctic lands have carried more ice in the recent past than they do at present: parts of Scandinavia are rising at rates measurable in centimetres per decade. Consequent changes to the coastline include successions of raised beaches, indicating that uplift was an intermittent rather than continuous process. In the Antarctic, the massive continental icecap has proved more stable and less liable to change than the several smaller Arctic icecaps. However, evidence of tectonic uplift appears in all coastal areas where exposed rocks and beaches occur, and is most marked on islands of the Scotia Arc, all of which have clearly been more heavily glaciated in the past.

Do such changes affect polar tourism management? Not directly, but they provide an opportunity for managers that is not found elsewhere, and should not be missed. Because of the sparse vegetation cover, nowhere in the world are the changes due to land-forming processes more vividly demonstrated than in polar regions. Major crustal structures show clearly: Iceland is one of the few points on Earth where visitors can walk into a major oceanic rift between tectonic plates, and see around them evidence of the forces involved. Rock strata, folding, fault lines, lava flows, frost and water erosion, formation of moraines, raised beaches and other periglacial features – all are clearly exposed, and need only a few words of explanation from a good presenter or guide leaflet to enrich the clients' experience.

Climatic variation during the ice age

Though temperate and tropical environments have featured continuously on Earth for hundreds of millions of years, polar environments are relatively new. Until about 20 million years ago, temperate conditions extended into both the present polar regions. Some five million years ago, Earth entered the current ice age (its first for over 250 million years) in which both polar regions gradually accumulated glaciers, at times exceeding the levels of glaciation that we see today (Box 4.2).

Box 4.2 Living in an ice age

World cooling toward the current ice age began over 60 million years ago, when Antarctica, formerly part of the Gondwana supercontinent, split from Australia and drifted slowly southward toward its present polar position. Fifteen to twenty million years ago, lateral pressures within the continental bloc, and later in the neighbouring mass of West Antarctica, had generated high, ice-capped mountain ranges. Circumpolar air and sea currents isolated the new continent from tropical and temperate regions, causing it to cool still further, and allowing the development of the huge icecap that has persisted since then. Earth's southern ice reflected back incoming radiation enough to lower temperatures throughout the rest of the world, causing the appearance some 2–3 million years ago of year-round ice in the north. Montane icecaps and glaciers spread to low ground, and the northern ocean became covered with permanent sea ice.

During the past 1–2 million years, the amount of ice covering northern lands and sea has fluctuated widely, due mainly to fluctuations in incident radiation. Though the concept of four major glacial events, based on classic studies in the European Alps, Scandinavia, Siberia and North America, is probably an over-simplification, during the last million years temperate and polar regions have passed through a dozen or more warm-cold cycles. Each cycle, lasting about 100,000 years, was marked by the build-up of ice in Greenland, northern Canada, Alaska, northern Europe and parts of Siberia, and its spread over neighbouring areas. Spreading was usually followed by a short period of stasis, then relatively rapid deglaciation. For a review of these findings and the evidence on which they are based, see Imbrie and Imbrie (1979).

During final phases of the most recent cycle, some 20,000–30,000 years ago, man first spread from tropical and temperate regions of Asia toward the Arctic coast, eventually giving rise to the many tribes and cultures that inhabit the Asian tundra, rivers and coastlands. Some 10,000 years ago, contingents of hunter-gatherers crossed from Asia to North America via Beringia, a broad lowland area since replaced by Bering Strait (Hopkins, 1967). Nine thousand years ago, conditions were about as warm as today, though more ice remained on land and sea. By this time, successive waves of human immigrants were spreading southward, eventually to the tip of South America. Others spread eastward across the Canadian archipelago to Greenland, founding the distinctive northern folk, formerly called Eskimo, now called Inuit in recognition of their pan-Arctic racial identity (Herbert, 1976; Brodie, 1987; Nuttall, 2004).

Six to seven thousand years ago, a warm spell, which climatologists called the 'climatic optimum', brought southern European forests to the

UK and dispersed ice altogether from the Arctic Ocean. Five thousand years ago, and again three thousand years ago, the cold returned, the pack ice re-formed, and the northlands were for long periods colder than at present. Within the past 2000 years, the Arctic has been subject to both cold and warm spells, with temperatures oscillating 1–2°C about the mean.

Warm and cold spells have been manifest in milder or harsher winters, more or less precipitation, longer- or shorter-lasting seasonal snow and ice cover. During warm periods, both land and sea ice were considerably reduced, glaciers retreated and perennial ice on both land and sea thinned or disappeared. As the treeline shifted north, forests invaded tundra, which in turn invaded polar desert. Tundra and desert were reduced in area but not eliminated: there remained refugia or 'safe areas' of original flora and fauna that expanded during later cold spells when the forests again retreated south (Lindsey, 1981; Haber, 1986: 59).

Again these changes have no direct effect on the management of polar tourism, but managers that do not provide information about them miss important points. Clients who visit polar regions usually do so because they are interested, not only in seeing the regions, but in understanding why they are different from the rest of the world. With fewer of the standard diversions of mass tourism available on call, polar regions offer opportunities for educational tourism at its best, a point discussed more fully in Chapter 5.

Recent climatic changes

Concern is currently expressed that normal global processes of climate have been seriously disrupted by human activities, in particular by the massive release of industrial gases (notably carbon dioxide and methane) into the atmosphere since the late 18th century start of the industrial revolution. Accumulation of these gases has enhanced the atmosphere's 'greenhouse' or blanketing effect, with the consequence that surface air and sea temperatures throughout the world are warming more rapidly than might otherwise be expected.

To examine the effects of global warming in the Arctic, the Arctic Council (representing the eight northern states with particular interests in the area) commissioned a comprehensive report on climate changes within the region in recent decades. The Arctic Climate Impact Assessment (ACIA, 2004) makes the following points (summarised in Stonehouse & Snyder, 2007: 38–39):

- worldwide climates are warming;
- recently, Arctic climates have risen twice as fast as in the rest of the world;
- this trend is likely to continue and accelerate during the present century.

Considering causes, the report concludes that the rapid warming is due mainly to the accumulation of greenhouse gases in the atmosphere. It also points out that the Arctic receives more ultraviolet radiation than in the recent past, due to depletion of atmospheric ozone, with less known but probably deleterious environmental effects. While regarding causes as important, the report is more directly concerned with the symptoms and consequences of change. These include:

- rising air and sea temperatures and increasing precipitation;
- shorter, warmer winters;
- widespread melting of glaciers and icecaps;
- reductions in snow and ice cover on land;
- thawing permafrost;
- flooding rivers;
- coastal erosion, crumbling land forms;
- reductions in extent, thickness and persistence of sea ice;
- changing strength and direction of ocean currents, notably the North Atlantic Drift;
- changing marine fish and wildlife distribution.

Poleward migration of the treeline involves loss of tundra, and possible invasions of competitive alien species and pathogens. Melting permafrost results in the destruction of roads and buildings. Diminution or loss of sea ice severely challenges ice-dependent marine mammals (notably seals and polar bears), planktonic organisms and dependent stocks of birds, fish and whales, and radically changes traditional hunting and fishing grounds on which indigenous human communities depend at least in part.

The ACIA report is particularly concerned that these trends are already discernable, that some are seriously affecting Arctic coastal communities, and that all will be considerable before the end of the present century. A range of sources confirm some existing effects.

- In Alaska, thawing permafrost, erosion and flooding directly affect 184 of 213 native villages (Government Accounting Office, 2003). The cost either to protect or to relocate those communities exceeds hundreds of millions of dollars (US Army Corps of Engineer Study, 2006). Other less tangible costs also need to be considered. As a native Alaskan commented on the potential disruption of 'lifestyles they've led for thousands of years that have been passed on to them by their forefathers. How can you minimize all that in economic terms?' (*Anchorage Daily News*, 10/22/07).
- In Greenland, rapid melting of the central icecap and peripheral glaciers, disruptions to animal life, changes in vegetation, loss of winter sea ice and amelioration of seasons are all implicated. Since

1995, the icecap has lost 7% of its mass and almost 100 m (300 ft) from its height. The coastline is altering: nunataks (in Inuit language 'lonely mountains') once locked into the icecap margins are being released to create new islands. Loss of winter sea ice is preventing dog-sledging, ice-hole fishing and seal hunting – the latter a source of traditional clothing, meat and oil – and similarly reducing polar bear hunting, once a cultural rite of passage. Estimates indicate that within the last decade the number of Inuit hunters has declined by more than half.

- In Canada, the native village of Salluit, Nunavik, near Hudson Strait, is becoming uninhabitable and may have to be relocated. Mean temperature of the permafrost has risen 1°C since 1988, mean air temperatures have risen since 1992 and the frozen ground is turning into mud (*Nunatsiaq News*, 2/29/08). Other Nunavut villagers are experiencing shortages of wildlife species that have for centuries provided sustenance and cultural identity, and irruptions of species not previously seen in this part of the Arctic, including robins, finches and dolphins (Bowermaster, 2007). Shorter winters and faster-melting sea ice reduce the annual number of hunting days.

In an address to the United Nations in April 2008, the Saami Council and Inuit Circumpolar Council stressed the need for quick response to the negative effects of climate change on native Arctic communities. The Saami of Arctic Norway, Sweden, Finland and Russian Federation, and the Inuit of Greenland, Arctic Canada and Chutkotka described the unique vulnerability of these people to climate change. Describing not only the need to address climate change 'in an urgent manner', they also emphasised that 'indigenous peoples, being the most marginalized on the planet, are most affected by the effects of climate change' (*Barents Observer*, 4/24/08). They point out that among these challenges, the two most daunting are:

- the direct impacts of climate change on their social, cultural and economic systems;
- the prospect that the Arctic will bear the burden of ill-conceived climate mitigation policies and programmes.

Again quoting from their address, 'The Saami and Inuit are highlighting the specific (climate change induced) challenges of the Arctic because they are real and hurt us' – a statement leaving no doubt that indigenous people feel immediate and direct threats from the changes that are happening in their environments.

In summary, the environmental changes now experienced by Arctic people, particularly indigenous people, are placing their cultures and subsistence economies at risk. This is the social, cultural and economic context within which Arctic tourism actually operates. What are the implications for the Arctic industry, on which many indigenous people are becoming increasingly dependent? This point is discussed more fully in Chapter 7.

There is no Antarctic equivalent to the Arctic Council, taking responsibility for generating similar comprehensive Antarctic studies. Nor, in the absence of an indigenous human population whose lifestyles are likely to be disrupted, is there a similar sense of urgency concerning changes in the Antarctic. Though losses of shelf ice from West Antarctica, losses of glacier ice from South Georgia and other Scotia Arc islands, and local changes in distribution and of sea ice have been reported, Antarctica's massive central icecap appears to stabilise local climates to an extent that is unknown in the Arctic.

However, changes in the distribution of plants and animals in Antarctic and sub-Antarctic ecosystems are to be expected: for reviews of existing ecology and possible changes, see Young (1991), Stonehouse (1991) and individual papers in Kerry and Hempel (1990).

Changes due to Human Activities

Though the remoteness and environmental hostility of polar regions have protected them against casual despoliation by man, there has been less protection against organised exploitation. Stonehouse and Snyder (2007: 35–36) list the following forms of human intrusion and exploitation that to one degree or another have affected polar environments:

- long-term use of the Arctic by indigenous populations;
- fur-trapping for distant markets from Arctic tundra and sub-Arctic forest-tundra;
- seal hunting for oil, furs and skins, and in the Arctic, for walrus ivory;
- whaling for oil and baleen in both regions;
- commercial fishing;
- extraction from the Arctic of minerals, including ores and hydrocarbons;
- establishment of military and scientific stations, refuges and other buildings intended for long-term use.

Each of these has resulted in some form of environmental change, from trivial to devastating.

Exploitation of living resources: Hunting, herding and fishing

Of the above list, least change appears to have resulted from long-term use of the Arctic's natural resources by indigenous populations. The hunter-gatherer Inuit of North America and Greenland found the region's meagre sources of catchable food and skins capable of supporting only small nomadic groups, who lived mainly on the coasts or rivers. Their exploitations were necessarily sustainable: had they not been, their cultures would not have survived. North American 'Indian' cultures, extending north to the Arctic fringe, were based largely on hunting and fishing. European and Asian native cultures came to depend also on migrating reindeer herds, which they followed and eventually husbanded. For an account of traditional subsistence hunting in Alaska and discussion of modern management problems, see Huntington (1992).

Reindeer herding continues throughout the Eurasian Arctic, providing meat, skins and a source of cash income for many communities with otherwise very limited resources (Vitebski, 2006). More prosperous communities with a wider range of income opportunities maintain contact with herding as an important link with heritage (Osherenko & Young, 1989: 87–89; Krupnik, 1993; Jernsletten & Klokov, 2002). Though many hectares of original tundra pasture have been contaminated by mining, smelting, development of hydro-electric power and other industrial activities, the industry is favoured by governments and remains strong. Visits to reindeer-herding camps are a growing tourist attraction in northern Scandinavia.

More damaging were the changes introduced by 18th- and 19th-century colonists from the south, who brought goods to trade for such local produce as furs and walrus ivory. Fur-trapping for southern markets was indeed the original motive for colonisation of much of the Arctic, notably Canada and Siberia, by southern states. Visitors from outside brought new aspirations, religions and education to Arctic and sub-Arctic communities, but also diseases, and demands on resources that were market-led and ultimately unsustainable. During two centuries of commercial fur-trapping, whaling and sealing for skins and oil, stocks of animals were severely reduced, to the detriment also of indigenous cultures.

Several species are currently recovering, but bowhead whales, once very plentiful, are still extremely rare, and subject to severe hunting restrictions. A few whales are taken annually under permit by high Arctic communities, mainly to maintain the traditional skills and social interactions involved. Whale-watching for tourists, guided by local naturalists, is becoming a preferred alternative. Juvenile seals are still taken for furs from parts of the Arctic pack ice, apparently on a sustainable basis.

Fishing, both freshwater and marine, has for centuries contributed to Arctic indigenous life. Commercial inshore marine fishing on a small scale has proved a viable industry for local communities, and deepwater fishing on a larger scale has brought prosperity to larger communities throughout the Arctic. Nuttall (1992) discusses aspects of changes in modern fishing and hunting practices, and their effects on communities of northeast Greenland.

The development of deep-water trawling and long-lining has opened both polar oceanic regions to commercial fishing. Attempts to control the industry by licensing are more successful in the Arctic, where policing is usually possible, than in Antarctic waters, where fishing grounds can seldom be fully protected. Changes in Arctic sea water temperature and salinity are resulting in losses or re-location of fish stocks, especially salmon, cod, sea run trout and herring, prompting all circumpolar nations to revise their commercial fishing regulations. Canadian, Norwegian and the US governments, for example, have shortened seasons, reduced catch limits, eliminated some commercial fishing for particular species altogether, and restricted the number of boats that can be engaged in commercial fishing operations.

To make use of expensive assets and sustain their seagoing way of life, Arctic residents forced to end commercial fishing have become charter boat operators to attract sport anglers. However, this transition depends entirely on the existence of healthy fisheries, and deep sea sport fishing too is now prohibited in many locations because of depressed fish stocks.

Similar restrictions and limitations are imposed on freshwater fisheries, including the lucrative sport fishing tourism market. Arctic rivers experiencing rising temperatures can no longer sustain salmon and other sea run species, requiring increasingly stringent regulations affecting types and abundance of species caught, length of seasons, allowable catches and catch-and-release regulations.

The Antarctic region, almost unknown before Cook's explorations of the 1770s (Beaglehole, 1974), was slower to yield its riches. However, sealing for furs and oil began on islands of the Scotia Arc before the end of the 18th century, and by the mid-19th century had almost completely destroyed stocks of both fur seals and elephant seals. No indigenous people were involved. Whaling began in the same area in 1904. After enormous annual harvests, first from among the islands, then from the periphery of the Southern Ocean pack ice, by 1965 stocks were severely reduced and the industry, by then in severe competition with vegetable oil production, was no longer viable. Antarctic seals and whales are now protected: cruise ships feature whale sightings as a tourist attraction, and record all whales encountered along their routes. Of the southern species previously hunted, humpback whale populations in particular appear to be recovering rapidly.

Trawling and long-lining fishing in Antarctic waters have developed since the 1960s. Though nominally controlled by the Commission for the Conservation of Antarctic Marine Living Resources (CCAMLR), an instrument of the Antarctic Treaty (p. 151), poaching is rife: several stocks are already in decline and others threatened. Long-lining takes not only fish, but a substantial by-catch of albatrosses, which has seriously eroded breeding populations throughout the southern ocean.

Exploitation of non-living resources

More damaging still were the ways in which non-living polar resources have been exploited, particularly in recent decades. In the Soviet Arctic, river diversions from the 1930s onward redirected immense quantities of fresh water from Siberian rivers to agricultural and industrial regions in the south, radically altering the balance of nature on the northern tundra and possibly affecting ice formation in the Arctic Ocean (Lamb, 1962, cited in Goldman, 1972). Water remaining in the rivers carries lethal mixtures of hazardous chemicals to the Arctic: Feshbach and Friendly (1992) report that the Angara River flowing northward from Lake Baikal to the Arctic Ocean has become:

> an aqueduct for poisons. Yearly it carries 257,000 tons of chlorides, 140,000 tons of sulfates, over 30,000 tons of organic wastes, and 10,000 tons of nitrates from factories built in the 1960s and 1970s along its banks.

Mineral ore and hydrocarbon extraction industries devastated large areas of the Russian Arctic from the 1920s onward. Zelikman (1989) records that most rivers flowing into the Barents Sea are heavily polluted due to nearby ore and chemical industries. For recent surveys of anthropogenic impacts on other Siberian Arctic environments, see Laletin *et al.* (2002) and Khitun and Rebritstaya (2002). Mineral extraction and related industries have similarly affected the North American Arctic, though generally on a smaller and more local scale.

Exploitation of mineral resources continues. The world's enormous demand for energy resources, coupled with evidence that the Arctic holds 25% of the world's hydrocarbon reserves, cannot fail to intensify development pressures on these resources throughout the Arctic. The Russian Federation is in direct competition with Saudi Arabia to be the world's largest producer of crude oil, and 70% of its production (7 million barrels per day) is pumped in wetland and permafrost environments in Arctic western Siberia. In Khanty-Mansi, the largest and most wealthy oil-producing region, 'the province's oil industry generates 40 billion dollars in annual tax revenue, 4.5 billion dollars of which Khanty-Mansi gets to keep for its own use. The rest goes to Moscow' (Starobin, 2008)

Massive oil and gas developments in Russian Arctic shelf seas currently cause environmental concern. Based on research in the Kara Sea, the Deputy Director of the Institute of Oceanology, Russian Academy of Sciences, comments that 'Unregulated navigation, active construction, and irrational mining may turn these sensitive systems into a heap of waste, which will be very hard to clean up' (Flint, 2008). Norway, the USA, Canada and Greenland are also actively exploring and making claims to potential hydrocarbon reserves located on their coastal shelves. International competition to secure ownership of the Arctic's coastal energy resources has caused sovereignty disputes (for a discussion of Norwegian/Russian disagreements, see Moe, 1991) that may be reconcilable within the context of the UN Convention on the Law of the Sea.

An unwelcome accompaniment to mineral extraction and processing is industrial haze, originating as smoke emissions from industrial plant. First noted in the 1950s, it has since been found to intensify every summer. Within the lowest 5 km of atmosphere, it reduces visibility, absorbs solar radiation and leaves measurable deposits of aerosol chemicals and other pollutants on snow, ground and vegetation surfaces: for details see individual papers in Stonehouse (1986).

Other sources of pollution throughout the Arctic are abandoned industrial, military, naval and scientific installations, from whaling factories to DEW-line stations. Often marked by downwind trails of refuse, occasionally containing such noxious materials as asbestos and explosives that are expensive and dangerous to remove, they stand uneasily between industrial archaeology and sordid rubbish. Such sites are, however, tourist attractions, as interesting and informative in their own way as more acceptable expedition huts and campsites.

Taking advantage of the ending of the Cold War in 1991, ministers of the eight Arctic states (Canada, Denmark (for Greenland and the Faroe Islands), Finland, Iceland, Norway, Russia, Sweden, the USA) came together to consider common environmental problems. Their particular concern was to protect the Arctic from further pollution, and where possible restore contaminated environments. A second concern was to encourage self-determination in indigenous peoples and promote their sustainable uses of natural resources. The long-term consequence of this initiative was formation of a permanent forum, the Arctic Council. A more immediate outcome was development of the Arctic Environmental Protection Strategy (AEPS), now a flourishing enterprise that identifies five major objectives:

- to protect Arctic ecosystems, including human societies;
- to protect, enhance and restore environmental quality, and encourage sustainable uses of natural resources by local communities;
- to encourage Arctic peoples in maintaining their traditional and cultural needs, values and practices, especially relating to Arctic environmental protection;
- to undertake regular reviews of Arctic environments;
- to identify, reduce and ultimately eliminate pollution.

To serve these objectives, four working groups were established – The Arctic Monitoring and Assessment Program (AMAP), Conservation of

Arctic Flora and Fauna (CAFF), Emergency Prevention, Preparedness and Response (EPPR) and Protection of the Arctic Marine Environment (PAME): for details of their functions, see Appendix B. AEPS is a clear and energetic response by the eight nations to the pollution accumulated over many years of industrial development. It recognises also the aspirations of indigenous people to play their part both in cleaning up the present and in shaping the future, which the Arctic Council is dedicated to encouraging.

However, development pressures and related environmental impacts are continuing, and may be expected to continue into the future. Wildlife and other natural features that attract Arctic tourists will continue to compete with the profits and jobs that industry – especially energy development – can generate. Is this a competition that tourism can hope to win? We discuss this point further below.

The Antarctic region, more remote than the Arctic from industrial and military imperatives, remains far less contaminated. Mineral extraction has never been attempted and is currently proscribed under Antarctic Treaty rules. Stations built and abandoned since WWII are being systematically removed – a welcome development after a generation of neglect. A few earlier buildings, including historic expedition huts and whaling stations, are now stabilised, tidied up and cherished.

However, despite its remoteness from civilisation, Antarctica has long been known to be contaminated by chemical insecticides, soil-dwelling pathogens and other pollutants derived mainly from atmospheric aerosols and marine currents, but also brought in by expeditions. As Stonehouse and Snyder report:

> Almost every corner of the continent and neighbouring islands has now been visited and to some degree contaminated by man. It would be difficult to sustain, in any technical sense, claims frequently made in official publications that Antarctica remains "pristine". It remains only relatively free from man-made pollution and damage. (Stonehouse & Snyder, 2007: 38)

This does not in any way detract from the natural beauty of Antarctica, the charms of its animal inhabitants or the need to protect it against further damage. The writers are concerned only at the use of a bad argument in a good cause.

Changes due to tourism

Tourism came to the Arctic in the mid-19th century and to the Antarctic a century later (Chapter 1). As an industry with an unenviable reputation for environmental damage, its operations in sensitive wilderness and semi-wilderness ecosystems, raises the question of whether tourism itself adds significant changes to those already affecting polar environments.

An annual influx of tens of thousands of visitors to areas that by definition are usually free of human influences cannot avoid bringing changes. Polar tourism is strongly seasonal: in simple terms, isolation and tranquillity depart with the arrival of the first tourist contingent in spring, and stay away until the end of the season. Scenery remains, and wildlife is remarkably forgiving, though even well-ordered tourism at either end of the world must interfere to some degree with both.

Despite the many forms that tourism takes, and the growing size of the industry, 'at neither end of the world is there evidence that tourism brings about environmental changes that match... natural or man-made changes' (Stonehouse & Snyder, 2007: 46). So, at least in the view of these authors, many of the qualities of remoteness and beauty for which polar regions are valued can be counted on to remain, even at the height of a busy season. The same authors, however, continue with a warning about the cumulative effects of natural changes upon indigenous people, and the additional strains imposed by tourism:

> The most severe tourist-induced changes occurring in the Arctic are probably those affecting indigenous human communities. Specifically, existing climatic and cultural changes impose risks on native cultures, and also bring benefits to them. An expanding tourism industry brings similar risks and benefits, which further complicate an already complex situation. The ability of the Arctic's indigenous people to survive has always depended on their capacity to adapt to change. Now, as they witness the changes wrought by reduced Arctic sea ice, and the influx of tourists that is at least partly consequential upon it, they face challenges that again put their cultures and livelihoods at serious risk. (Stonehouse & Snyder, 2007)

Should indigenous people of the Arctic be forced to abandon even some of their communities through tourism, fewer opportunities will exist for tourists to appreciate their cultural traditions, or for tour operators to prosper. It is thus in everyone's interests that visits are managed in ways that impose least upon the hosts.

Subsistence economies of indigenous peoples throughout the Arctic are particularly vulnerable to environmental change. Traditional dependence on hunting, herding and fishing is now seriously threatened, as are many of the small-scale industries that are their modern equivalent. Existing changes occurring in wildlife habitats are reducing the availability of seals and polar bears, altering the migratory patterns of such other hunted species as caribou, and changing or eliminating seasonal fisheries. Further threats arise from the Arctic's energy development boom, which exerts competitive pressures on critical wildlife habitats and increasingly scarce wildlife species. Expected changes will further reduce opportunities for the old ways of life – all adversely affecting subsistence economies and the cultural integrity of the Arctic's native people. These points are discussed more fully in Chapter 7.

A comprehensive pan-Arctic assessment of recent changes in human communities and their environments is provided by the *Arctic Human Development Report* (Larsen, 2004). Compiled under the auspices of the Arctic Council's Sustainable Development Working Group to 'provide an integrated picture of the state of human development in the circumpolar Arctic' (Young & Einarsson, 2004: 15), this includes 11 wide-ranging chapters covering demography, socio-cultural, political and economic systems, new and sustainable approaches to harvesting renewable resources, viability of communities under pressure from change, health, education, gender issues and international relations. Together these indicate how radically Arctic human systems have changed in recent decades, dispelling any possible misconceptions that 'Nanook of the North' still typifies Arctic culture, but stressing that groups of indigenous folk in remote communities remain at least partly dependant on local renewable resources, and thus vulnerable to the environmental changes that are currently happening (Ford *et al.*, 2006).

Many native peoples throughout the Arctic, particularly those recently attaining self-rule, view tourism as economically more reliable than tenuous dependence on either subsistence or resource extraction economies, and are developing heritage tourism programmes. Examples are the Alaska Native Corporations, the Saami throughout Scandinavia, the Inuit of Nunavut and of Greenland, the people of Iceland and the 42 indigenous groups comprising the Russian Association of Indigenous Peoples of the North (RAIPON, 2007). All believe that tourism represents a means for acquiring jobs and income as well as creating markets for native products and services, and are determining how their culture can be shared by means of tourism enterprises. They perceive too that thoughtfully designed heritage tourism programmes provide methods for preserving traditional ways of life, language and cultural values (Amberger, 2003; Milne *et al.*, 1995).

Tourist experiences may include participation in selected activities, or be limited to the purchase of goods and select services (Alaska Native Council, 2005; Walle, 1993). Participatory tourism has both advantages and drawbacks. Each year, when the Arctic's permanent residents are invaded by huge numbers of visitors, determining how to share their culture and natural resources involves the hosts in difficult choices. Protection of privacy has to be balanced with allowing entry to communities, traditions and ceremonial practices. Insensitive acts by tourists, or even their very presence, can stress the ways of life and cultures of people who live remote from outside company for much of the year. Such stresses can only increase as tourism continues to grow.

Tourism also requires facilities and infrastructure that are expensive to provide and maintain, placing a significant economic burden on governments or private entrepreneurs that seek to encourage it. The economic benefits that it brings appear during a brief season each year, but may have to sustain households throughout the year. Those who host tourism are vulnerable to economic disruptions, adversely affecting the industry, which are entirely beyond their control.

However, advantages currently appear to outweigh disadvantages. Arctic economies are increasingly dependent on tourism for jobs, household income, business and tax revenues and capital investment. For Arctic resource managers and policy makers, key issues are:

- identifying and measuring environmental impacts that are caused by tourism;
- identifying benchmarks for monitoring environmental conditions;
- determining levels of resource uses that will not stress local subsistence economies;
- creating trade-offs between non-renewable and renewable use of natural resources;
- identifying, assessing and minimising cumulative impacts.

These tasks are challenging for Arctic policy makers, even more so for managers who face the day-to-day problems of accommodating an ebullient and growing industry. For example, management decisions regarding cruise ship transits, allowable entry to marine-protected areas, critical fisheries and critical wildlife zones require environmental protection policies and techniques. But they also require knowledge of the rôle of climatically induced changes, consideration of the economic dependence of local communities on tourism, respect for the desires of native people to approach self-sufficiency through tourism, and the need to protect the safety of tourists in the Arctic.

Changing Sovereignty Issues

Canada, Denmark (on behalf of Greenland), Norway, Russia and the USA are the nations with most direct coastal interests on the Arctic Ocean. Reductions in Arctic sea ice extent, duration and thickness, together with rapidly increasing world demand for energy resources, have set in motion attempts by all these nations to re-define their jurisdictional boundaries in the ocean. Under the 1982 United Nations Convention on the Law of the Sea (UNCLOS), coastal states may in certain circumstances claim ownership of seabed beyond an existing 200 nautical mile (370 km) limit, to one of up to 350 nautical miles (648 km). Should claims to extended sovereignty involve disputes, UNCLOS has provision for arbitration. However, changes in jurisdictional authority are likely to affect both the allowable uses by tourists of environmental attractions, and the operational guidelines under which they are licensed. For further details of UNCLOS, see Chapter 5 and Appendix A: for an account of its development and principles, see Koh (1991).

These claims are to secure title to as-yet undiscovered oil and gas reserves. One most highly publicised recent claim was Russia's, in planting its flag on the seabed at the North Pole. Less showy claims have been pursued by Canada and the USA. For example, Ottawa has recently announced that it will double its spending on scientific research projects to $40 million over four years in an attempt to prove that the North American continental shelf extends far beyond Canada's 200-nautical mile limit. If successful, Canada may legally claim jurisdiction over an extensive submarine area of promising oil and gas reserves (Canada.com, 5/14/08).

The coastal states have until 2013 to make their case to UNCLOS for sovereignty over the continental shelves. Given evidence that all claims are likely to be contested, Demark hosted a conference in Ilusaat in May 2008, to begin negotiations for resolving disputes. Of a variety of solutions proposed, none was accepted. Other matters discussed included maritime safety, environmental safeguards and incidence response – all issues that will inevitably affect Arctic tourism management. The Arctic marine environment, once a domain within which all ships could move freely, will almost certainly experience new governance and enforcement mechanisms.

A similar situation is likely to arise in Antarctica, where parts of the (much deeper) continental shelf may contain hydrocarbon deposits. There the issue will be complicated by questions of ownership, which participation in the Antarctic Treaty has only partly resolved. Can Australia, for example, expect its claim to part of the seabed surrounding the continent to be taken seriously, while its claim to part of the continent itself is recognised by only a small minority of other nations? (Rothwell & Scott, 2007: 15–16).

Summary and Conclusions

This chapter discusses polar tourism in the context of environmental changes. While we are constantly reminded that warming climates are bringing changes to the world, including the polar regions, many other forms of change have occurred, and are still occurring, and need to be taken into account in effective tourism management. We distinguish two categories of changes: those due primarily to cosmic events, some of which may be intensified by human activities, and those due primarily to human activities, including hunting, prospecting, mining, landscape modification and tourism itself.

Changes within the first category are usually beyond human control: we cannot alter the movements of crustal plates, and the degree to which we may hope to control climatic change is a subject for lively discussion elsewhere. Yet both influence polar tourism, one by affording spectacular attractions (e.g. the volcanoes of Kamchatka and southern Alaska in the north, Deception Island, the South Sandwich Islands and Mount Erebus in the south), and the other by opening up new routes. Both provide educational interest to enhance the tourist experience. Changes due to human activities are more tangible, often to the detriment of an ideal Arctic and Antarctic, but nevertheless to be taken into account as part of the polar scene.

We discuss past and present uses of both renewable and non-renewable resources, and outline the current rôle of tourism as a contributor to the welfare of Arctic indigenous communities – a point to be raised further in Chapter 7.

Chapter 5

Wilderness Tourism: Challenges and Techniques

Introduction: Wilderness Challenges

Though subject to many definitions, 'wilderness' is generally agreed to be large expanses of land that are naturally wild, primitive, remote, virtually undeveloped and generally shunned by man (see: **Wilderness area management**, p. 82). So defined, many of Earth's largest wildernesses occur in polar regions. The Antarctic continent is the world's largest single wilderness, and enormous tracts of Arctic land, though long inhabited by man, retain a credible wilderness character. Polar wilderness regions are extensive ecological zones that have sustained their environmental integrity for millennia, defended and kept intact originally by their isolation, more recently by the difficulties and expense of gaining access (Figure 5.1).

Throughout most of human history, wilderness regions were deemed hostile and dangerous places that were best avoided. Those fears were well substantiated by the wild animals and warring peoples who inhabited them and by the lack of arable land to afford sustenance. They were, for millennia, places to be tamed and conquered. The mantra of 'Go west, young man' was an invocation to seek wilderness, conquer it and replace it with farmland. Given that prevalent attitude, wilderness regions were subjected to radical change that transformed them from inhospitable regions to safe and economically productive lands. Despite that extensive history and its associated land use practices, it is remarkable that people now laud the value of wilderness and protect their wild character and dynamic environments.

New concepts regarding wilderness management emerged from the late 19th century onward, especially in the USA where the need to protect wilderness was first recognised as an important issue. These concepts culminated in the creation of 'wilderness management techniques' that seek to preserve rather than alter the fundamental character of vast regions. The fact that wilderness – naturally wild places – needs to be managed may at first sight seem a contradiction in terms, but wilderness throughout the world is encroached on by man, and needs managing to retain its wilderness qualities. Today, remaining areas of wilderness are cherished, and wilderness management is the business of using wild places for non-destructive purposes that retain their essential wildness.

Figure 5.1 Aircraft provide access to remote wilderness areas in both polar regions. Photo: JMS.

Not surprisingly, much current thinking on polar management at either end of the world is based on US models.

Despite their remoteness and severity, polar regions are attractive to competing interests that seek to realise value from them. Developers look to extract energy resources and minerals, commercial fisheries search for new fish stocks. Environmentalists look for ways of conserving and protecting their integrity. Tourists seek to experience their wildness, and enjoy their scenery, plants and animals. Arctic governments generally encourage all these approaches, hoping to draw revenues from wilderness areas that will help to defray management costs, and provide permanent residents of wilderness with paid employment.

Balancing economic development and environmental integrity in these areas is predominantly the responsibility of governments, who must take into account available finances, the need for physical infrastructure and the proximity of human resources. Approaches for the Arctic and Antarctic are quite different, though equally difficult to accomplish. Immediate challenges are economic development pressures, immensity of scale, severe environmental conditions, difficulty of access and the lack of either physical or human infrastructure to undertake basic management functions or respond to emergencies – all circumstances that conspire against easy, inexpensive approaches to management.

This chapter explores management responses to these challenges and describes both a general framework and specific techniques that have been applied to managing wilderness tourism in the Arctic. Antarctic wilderness management options are discussed in more detail in Chapter 8.

Economic Development in Polar Regions

Close enough to centres of population to be exploited, the Arctic has for centuries been susceptible to development for profit (Chapter 1). Primarily because of its resources (notably its mineral wealth), secondarily to meet the needs of its human populations, development is currently encouraged by all Arctic governments. Jurisdictions, agencies and organisations compete over alternative uses for its resources; Arctic managers are under pressure to balance competing interests, and to create and enforce regulations that address a diversity of wilderness uses. Among the contestants are tour operators, competing for use of those resources that their clients particularly value.

The Arctic is governed by eight sovereign nations (p. 72), each of which has authority to decide how resources within its domain are used. Each nation is responsible for communities of indigenous people with traditional skills and practices, who have simultaneously used and conserved local resources according to their own cultures (Chapter 7). The eight nations work independently of each other, setting their own objectives, standards and regulations. Faced with common management problems, they cooperate through the Arctic Council in setting common objectives and encouraging what they jointly determine to be allowable uses.

In contrast to the Arctic, Antarctica is governed by the Antarctic Treaty System, controlled by a consortium of nations that has no ambitions to develop Antarctica in ways in which the Arctic has been developed. Though the continent has its share of mineral wealth, including hydrocarbons, the presence of a huge icecap ashore and some of the world's largest icebergs in surrounding waters restrict possibilities of extraction. There is no indigenous population whose interests would be served by development. The Treaty powers have banned mineral development and limited possibilities of sealing: the seas around are whaling sanctuaries. Scientific research is established as the highest priority use of the continent. Under the Treaty's Convention on the Conservation of Antarctic Marine Living Resources (CCAMLR), attempts are being made to control Southern Ocean fishing by licensing – an objective frustrated by high-technology poachers in an ocean that has so far proved too big to be patrolled effectively.

Environmental management south of 60°S is effected in terms of the Protocol on Environmental Protection to the Antarctic Treaty, an instrument of the Treaty that came into force in 1991. Tourism, the region's newest industry, has been recognised as a legitimate activity, but is so far controlled only under the general terms of the Protocol: further details of this management system appear in Chapter 8. The system lacks two important management features that are available in the Arctic:

- there is no indigenous population with direct interests in caring for the environment;
- there is no money from revenues to carry out even the smallest management functions.

Wilderness management in both polar regions involves balancing increasing human activities with the need to conserve natural (and in the Arctic, heritage) resources that are already subjected to harsh conditions in changing environments. Ideally, management objectives should be stated by Arctic governments and Antarctic Treaty parties in terms of responsible stewardship and sustainability. The prevailing reality normally involves contentious debating among a diversity of stakeholders to determine allowable uses. The Arctic benefits from its nations' ability to exert jurisdictional authority and dedicate resources to wilderness management objectives. The Antarctic is managed by a regime that, cumbersome and slow moving, is still experimenting tentatively with the realities of resource management, lacks funds to create on-site management regimes, and has neither money nor legal powers to enforce its decisions.

Wilderness Area Management

Wilderness area legislation is a useful source of information on tourism management criteria, because it already deals with recreational uses. By example, for two decades following WWII, advocates of wilderness legislation in the USA were concerned with far more than geography, natural attractions or physical dimensions. They defined wilderness in such philosophical terms as places of solitude and personal rejuvenation, possessing spiritual values that required protection. Proponents of wilderness area designation, such as Robert Marshall, Aldo Leopold, William O. Douglas and David Bower, sought to establish large expanses of public wilderness habitat that would provide for invigorating physical and spiritual experience, to the exclusion of destructive commercialism (Wellman, 1987).

Leaders of the wilderness legislation movement were also politically astute enough to realise that some economic value needed to be identified in order to enlist public support, and offset the huge clout of pro-development economic interests. Recreational use of wilderness areas, especially low-impact recreation, provided means both for bringing in revenue and attracting the political support needed to pass legislation (McCloskey, 1970). The convergence of these purposes and motives was codified in the 1964 Wilderness Act, under which wilderness was defined as:

> land retaining its primeval character and influence, without permanent improvements of human habitation... protected and managed to preserve its natural conditions. These areas also allow ...outstanding opportunities for solitude or a primitive and unconfined type of recreation. (PL 88-577, Section 2c, quoted in Hendee *et al.*, 1990)

Wilderness legislation proclaimed that recreation was one of the 'highest and best' economic uses for wilderness areas, a rationale supported by arguments that recreational use was a renewable, long-term use of resources. As such, it was economically preferable to the alternative short-term, non-renewable, exploitive uses, e.g. timber harvesting, energy development and mineral extraction. That argument was further reinforced by the declaration that wilderness was an increasingly scarce resource and thus gained economic value through time.

Recreation and protecting the environmental integrity of large regions became the primary purposes and thus the most important allowable uses for wilderness areas. Simultaneously, legislative initiatives to establish a Wilderness Area System in the USA were aided by the public's growing concern for nature that would evolve into what is now called the 'environmental movement'. The juxtaposition of these people, motives and events initiated the creation of protected areas, not only in the USA, but throughout the world. From the need to effectively manage such areas arose the management objectives and methods that are widely used today.

During the last century, jurisdictions throughout the Arctic have officially designated vast tracts of land and sea as Wilderness Areas, Marine Sanctuaries, National Parks, Wild and Scenic Rivers, and Wildlife Refuges. Collectively, these designations represent substantial commitment to the conservation of the wilderness character of the Arctic (see Appendix B).

Tourism Management Objectives

Tourists visiting the world's wild places by ship, aircraft or coach present a challenge to natural areas that may be particularly sensitive to disturbance. Wilderness is by definition a place with few or no people: that is one of its attractions to visitors from the towns and cities of a crowded planet. Thus, visitors *en masse* are likely to alter what they have come to experience. If there are too many of them, and they behave insensitively, alteration can quickly become damage that, if not entirely reversible, takes time for natural processes to repair. Nowhere is this more true than in polar regions.

Is the remedy, then, to keep tourists out of wilderness areas, and especially from polar regions? No: even if desirable, that would be impossibly expensive. Who would pay to keep people out of deserts? Excluding people arbitrarily from wilderness would be questionable democracy and worse conservation practice, for wilderness needs managing, management costs money, and nobody invests money in places where the warmest invitation is 'Keep out'.

Instead, as a world community, we laud and encourage tourism, including visits to wilderness areas, relying on sound management and conservation measures to keep tourists safe and the environment intact. That is what good tourism should be about. But what are the rules of management?

There is no single manual covering tourism management and conservation in wilderness areas, nor is one likely because of the incredibly diverse ecologies they contain. The principle goal of managing the environmental integrity of wilderness means that each management plan has to be specially suited to its unique environmental conditions. Competent management thus requires a competent understanding of the environmental factors that sustain the primitive character of a wilderness area. Gaining that understanding, as a prerequisite for management, is admittedly a tremendous challenge. But the preservation of a region's wilderness character demands that type of research effort.

Wilderness includes Earth's hottest and coldest landscapes; it may or may not support indigenous populations, and compared with more hospitable regions, it attracts visitors only in small numbers, thereby generating little in the way of revenues. When tourism comes initially to wild places, governing authorities are likely to provide policy statements and codes of practice rather than regulations, outlining management features that tour operators are expected to meet. More stringent regulation may follow when the tourism is better established and has started to pay for its privileges.

Criteria for Polar Tourism Management

The *Concise Oxford Dictionary* (9th edition: 1996) defines 'manage' as 'to organise, regulate, be in charge of'. 'Management', the resulting process, thus implies organisation for a particular purpose within a framework of regulation and authority. In any field of endeavour, organisation is based on objectives stated and intended to achieve a particular purpose. Without clear objectives, both organisation and purpose are lost, and management founders.

Accepting these precepts, sound management of tourism or any other exploitive activity requires:

- a management organisation with clear objectives, operating within –
- a regulatory framework provided and maintained by an authority with knowledge of, and strong commitment to, sustainable management of resources.

In tourism, the primary management organisations are most often tour operators, and the regulatory frameworks are (or at least should be) provided by governing bodies responsible for the resources on which tourism draws. But the governing bodies themselves are usually hierarchies involving further management systems, e.g. national park authorities, wardens, rangers and licensed guides, on which direct responsibility for the resources devolves. Those too require organisation, objectives and terms of reference to discharge their duties properly.

There is a vast literature on tourism management in general, with a small but growing proportion concerning polar tourism. Can we gather from it what characteristics or properties of management apply especially to polar tourism?

Sustainability, responsibility

'Sustainability' and 'social responsibility' appear often in the literature as key concepts in tourism management. Johnston and Hall, for example, cite one and imply the other in their summing-up of a symposium volume on the management of polar tourism:

> there is an urgency in formulating sustainable approaches to tourism at the poles. Sustainable tourism means conserving the productive basis of the physical environment by preserving the integrity of the biota, ecological processes and cultural values, and at the same time, producing tourism commodities without destroying other aspects of land use such as indigenous peoples' activities. (Johnston & Hall, 1995: 309)

This implies that, though the industry is growing steadily and urgent action is needed, sustainability remains an all-important criterion for management. Similarly, social responsibility is a keynote: the rights of tourism must be fostered, but not at the expense either of the environment or of other legitimate users.

Though widely used, generally agreed to be desirable and readily defined in imprecise terms, 'sustainable tourism', 'responsible tourism' and similar values cited in management primers are difficult both to deliver and to assess. There is an assumption that everyone knows what they mean, but because their objectives are seldom defined, they cannot be implemented by management except in general terms.

Sustainable development came to prominence through the International Union for Conservation of Nature and Natural Resources (IUCN) World Conservation Strategy (IUCN, 1980) and was encapsulated in a memorable definition of the subsequent Brundtland Commission Report, Our Common Future:

> sustainable development is development that meets the needs of the present without compromising the ability of future generations to meet their own needs. (UN World Commission on Environment and Development, 1987: 43)

Thus, sustainability in tourism can be said to be achieved if the objectives for which tourists travel are not destroyed by their visits – if scenery, wildlife, cultures and other resources remain unspoilt, indeed unaltered, to be used repeatedly in the future.

Tourism worldwide presents many examples of the business that have not proved sustainable, because the resources on which they were originally based have been marred or ruined (see, e.g. Boo, 1990; Butler, 1991; McKercher, 1993). This does not necessarily mean that visits have stopped – only that some alternative purpose has been found for maintaining them. Especially vulnerable are wild places, which are visited at least in part for the sake of their wildness and isolation. Wildness and isolation are readily destroyed by too many visits. Once they are lost, there may be no alternative uses.

Why should a condition that is so simple to state, and so easy to understand, prove so elusive in practice? A further definition later in the Brundtland report provides one possible answer – sustainability is:

> a process of change in which the exploitation of resources, the direction of investments, the orientation of technological development, and institutional change are all in harmony and enhance both current and future potential to meet human needs and aspirations. (UN World Commission on Environment and Development, 1987: 47)

This warns in effect that the industrial world deals in money, technology and human attitudes, all of which must combine to agree on development policies. When the tourist industry budgets for these factors as well as for environmental welfare, there are no built-in guarantees that harmony will prevail and the resulting tourism be sustainable.

Of many circumstances in which these unintended consequences occur, two in particular are worth mentioning in the context of polar tourism.

- The first involves the use of tourist operational best practices without consideration for the cumulative impacts associated with those practices. Specifically, a few tour groups implementing minimal impact approaches to wildlife viewing and resource conservation may have no measurable negative impacts on an environment. But the cumulative impacts resulting from many groups visiting exactly the same area may produce significant adverse effects on animal behaviour and habitat conditions. In these circumstances, each tour operator may employ valid management techniques, but in vain if too many tour groups are visiting a site.
- The second set of circumstances is the use of operational best practices that are suitable for most of the year, but inappropriate during special seasons. Seasonal dynamics, such as nesting, breeding and foraging, which uniquely characterise an environment, can be disrupted by visitor intrusion that would otherwise not be harmful. Nowhere is this more true than in polar regions, where seasons are particularly marked and environmentally critical.

Considering Brundtland's concept in relation to the Arctic, Griffiths and Young are also cautious, regarding sustainable development as a contradiction in terms:

> While it may be the notion currently available to guide us toward a decent life in a habitable natural environment, it is open to contrasting, sometimes conflicting and, in our view, frequently misguided interpretations. In the hands of politicians, officials business leaders and representative of the media, sustainable development is coming to mean *sustainable economic growth* – enjoying "all the growth you want so long as it does not destroy the biosphere" or doing a more careful job of what has brought us to our present impasse. Understood in this way, it is no more than a palliative, comforting perhaps in the short run but incapable of curing the basic problem in the long run. (Griffiths & Young, 1991: 33)

These authors see sustainable development as no more than a 'fashionable catchword' (Griffiths & Young, 1991: 35), one that 'sounds a welcome note of pragmatism and conveys a message of hope to a world increasingly burdened with anxieties about the future'. They regard it as impossible to achieve among small, scattered Arctic communities, and argue instead for another form of development, not in industry but in humanity itself – one derived from examples set by the indigenous people of the Arctic:

> a cultural revolution – fundamental changes of thought and behaviour allowing humanity to create social and natural preconditions for an existence that respects and adapts to the natural environment.

Considering tourism in the Canadian Arctic Northwest Territories, Butler, too, regrets that 'sustainable development' has become a buzz-word, and needs careful definition in management terms. For tourism to be truly sustainable, it must be integrated with the physical and cultural environment and other economic activities in the area in which it operates:

> To achieve sustainability, two key factors need to be considered, the ability to control tourism and the capability of the area to withstand use. With respect to the first point, there must be:
>
> 1 a desire to control tourism,
> 2 a policy or strategy to achieve this,
> 3 institutional arrangements or means to implement the strategy, and
> 4 opportunity to do so.
>
> In the context of the area itself, we need to consider:
>
> 1 The environmental characteristics and quality.
> 2 The human/cultural components, their degree of development and economic dependence.
> 3 The wildlife and its ability to withstand disturbance and consumption.
> 4 The geographic patterns of elements such as tourism, local settlements and ecological features. (Butler, 1992: 18)

Butler's first four points imply firm governance, which is often present on paper but missing in the field, particularly from wilderness areas. This is not surprising: areas that are difficult for operators and tourists to reach are also difficult and expensive to administer, often yielding small returns in revenues and fees to support their governance. However, 'responsible tourism' depends on those in charge accepting responsibility, equipping themselves with sensible regulations that they are prepared to enforce, and standing close enough to the industry to be effective in monitoring and control. Those who claim to govern tourism have key responsibilities that cannot be surrendered to tour operators, even less to the tourists themselves.

Butler concludes:

> One has to be somewhat pessimistic with respect to the probability of achieving tourism in a sustainable development context in the Canadian Arctic. Individually it is likely that all players in tourism would accept that principle as appropriate and desirable, but implementing it is another matter... There is... the real need to place effective control over both external and internal elements, and to set clearly defined and generally accepted goals for tourism in appropriate locations. (Butler, 1992: 19)

Butler's pessimism is germane particularly to the desire of Canada's native people to determine how their homeland is used. Having recently attained sovereignty, they are simultaneously the most knowledgeable stewards of their resources, and the ones most responsible for the economic development of tourism in their corner of the Canadian Arctic. Determining how they will host tourism is not a decision to be taken lightly (Dressler, 2001).

When considering realistic approaches to managing tourism, the discouraging pronouncements of some researchers need to be tempered. Much of the tourism literature focuses exclusively on tourist numbers and their control. A far more realistic approach to tourism management distinguishes alternative tourism experiences. This approach recognises that what, where, how and the season when tourism occurs represents the true tourism management issues that need to be addressed. This approach more accurately identifies the probable impacts that tourism has on polar environments and cultures.

Tourist numbers are important, but what tourists are doing and where they are doing it is substantially more important. The thousands of tourists who browse the shops of Arctic towns are less of an environmental threat than the hundreds who are rafting, kayaking or backcountry camping in remote polar regions. Similarly, the lone sailor with friends and family is a perpetual concern to emergency responders who normally have no idea of the wanderers' existence until a calamity occurs. Truly effective tourism management requires the selection of techniques that are uniquely adapted to specific recreation activities and the environmental conditions where they are pursued. Crafting environmental and tourism safeguards and reinforcing them with regulations, monitoring and approved tour operator procedures are works in progress throughout the polar world.

These views indicate that 'sustainable development' and its derivatives 'sustainable tourism' and 'responsible tourism', though widely used and with popular and comforting overtones, in themselves represent only aspirations. We discuss below some of the steps needed to convert them into achievements.

Environmental integrity

With the development of cheap mass-travel during the 1970s and 1980s, tourism worldwide expanded into exotic venues, many of them in wild environments that had not previously been visited, including the polar regions. Ecologists and conservationists expressed concern for the future of these environments, which they rightly pointed out were unused to human invasion and could quickly be ruined.

Small-scale operators, who for long had led parties into out-of-the-way places, with minimal environmental impacts, found themselves in a limelight that might have destroyed their market. Instead, environmental concern worked to their advantage: the market for small tours to wild places took off, under such names as 'nature tourism', 'adventure tourism', 'alternative tourism' and 'ecotourism'. New operators were quick to join in. Though many operators fell by the wayside (due mainly to lack of business acumen: McKercher, 2001: 566), this lively market segment continued to attract clients looking for experiences beyond the commonplace, with its names indicating departures from the banalities of mass-tourism toward more laudable objectives.

Conservationists who had hoped to see fewer tourists visiting wild places were disappointed. In compensation, the new wave helped to popularise an ethic of environmentally responsible tourism that has since permeated the whole industry.

In newly established tourism sectors, operators and clients alike are on their best behaviour, treading warily to avoid environmental degradation. In well-established venues, less care may be taken, short-cuts developed, more economic and cost-effective measures introduced, safeguards scrapped – all increasing the chances of a decline in value of the resources that were formerly treasured. The remedies are a careful initial assessment of the resources, a management plan with clearly stated objectives and periodic monitoring to ensure that the objectives are maintained.

Ecotourism

Prominent among forms of nature-based tourism that evolved during this period, ecotourism was defined by Ceballos-Lascurain as:

> travelling to relatively undisturbed or uncontaminated natural areas with the specific objective of studying, admiring, and enjoying the scenery and its wild plants and animals, as well as any existing cultural manifestations (both past and present) found in these areas. (Ceballos-Lascurain, 1987: 14)

In its earliest manifestations ecotourism was virtually synonymous with 'nature tourism'. More recently, it has developed into a branch of tourism in its own right, with an overlay of separate objectives that have been reviewed, elaborated, argued-over, re-defined many times, and indeed been made the subject of their own *Encyclopaedia of Ecotourism* (Weaver, 2001). Blamey (2001: 6) sums up its essential qualities under three simple headings; ecotourism is:

- nature based;
- environmentally educated;
- sustainably managed.

Wight (1994, quoted in Blamey, 2001: 11) presents a longer list of principles that take ecotourism well beyond Ceballos-Lascurain's definition. In addition to developing in environmentally sound ways that do not degrade its resource, ecotourism should also:

- provide long-term benefits to the resource, the local community and industry;
- provide first-hand participatory and enlightening experiences;
- involve education among all parties, including local communities, government, non-governmental organisations (NGOs), industry and tourists (before, during and after the trip);
- encourage all-party recognition of the intrinsic value of the resource;
- involve acceptance of the resource on its own terms and in recognition of its own limits, which involves supply-oriented management;
- promote understanding and involve partnerships between many players, which could involve government, NGOs, industry, scientists and locals (both before and during operations);
- promote moral and ethical responsibilities and behaviour towards the natural and cultural environment by all players.

Honey (1999: 19) advocates ecotourism as a better use of a poor country's natural resources than logging and agriculture. Many other authors add their own choice of qualities, generating what could easily be mistaken for wish-lists of ideal tourism, rather than prescriptions for the sound management of an industry. This does not in itself devalue ecotourism; though it may lead to suspicions that the term has become a marketing gimmick as much as a statement of principles.

Buckley (2001: 379 *et. seq.*) sums it up as a development of nature tourism, involving progressive improvements that above all minimise its environmental impacts:

> ecotourism should involve deliberate steps to minimize impacts, through choice of activity, equipment, location and timing; group size; education and training; and operational environmental management. ...the impacts of ecotourism should therefore be those of nature tourism and recreation which involves best-practice environmental management

McKercher (2001: 569) is in no doubt that the attractiveness of the term has been abused by marketers and promotion agencies alike, with the result that almost any form of non-urban tourism can now be called ecotourism. However, he represents the views of many colleagues in pointing out that:

> Ecotourism is widely recognised as an ecologically, morally and ethically preferred form of tourism that, if done correctly, optimizes

> social, cultural and ecological benefits, while providing the tourist with an uplifting experience. (McKercher, 2001: 565)

While 'ecotourism' may be suspect as a label, the principles behind it remain sound. Issaverdis (2001: 579) proposes four remedies to combat misrepresentation, which have proved practicable and successful in the wide expanses of Australia:

- Benchmarking: selecting target features and practices, and monitoring to see that they are achieved.
- Accreditation: establishing consistent industry-nominated managing standards and practices, and accrediting operations that meet them.
- Setting standards in best-practice business criteria.
- Auditing, frequently carried out to confirm and update benchmarks, accreditation and best-practice standards.

For environmental implications of these points, see **Best-practice management** below. Issaverdis and other authors who support ecotourism principles stress that operational success rests in good forward planning, and the responsibilities of setting, maintaining and monitoring standards being shared between tour operators and regulating authorities, taking into account the views of all other interested parties. So a harmonious outcome can of course be achieved without being called 'ecotourism'. But the term becomes useful where tour operators tend to work *ad libitum*, regulating authorities are lax or uncertain of their responsibilities and environmental degradation is apparent – an unhappy combination that all too often applies in wild areas, including polar regions. If a more principled methodology is required, the principles of ecotourism make a sound starting point.

Biodiversity, bio-integrity

If the principles of ecotourism are right for protecting wild areas, applying them in tourism management requires careful thought about objectives. If benchmarking, accreditation, standards and auditing are to be designed and put into effect, what are the key objectives to which they relate?

Sustainability is a key objective in any tourism operation: responsible operators want to be able to repeat their successful ventures year after year. So they need to conserve the environment in which they are working. What must they seek to protect, and how will they know if they are succeeding? What are the benchmarks and standards?

To maintain sustainability, it is important to minimise impacts of tourists on the environment: so much is generally agreed. But as Minbashian (1997: 18) points out, 'minimising impact' alone is a poor management goal. It emphasises what we do not want without telling us the characteristics or state of the system that need to be conserved, and offers no guidelines for positive action.

As the attractions of wild places are often predominantly biological, it is important not to destroy the integrity or wholeness of whatever biological systems are evident. The distinguished US ecologist Aldo Leopold (1949: 224–225) noted that, in ecological terms, 'a thing is right when it tends to preserve the integrity of the biotic community'. On similar lines, and arguing for a positive approach, Minbashian suggests that a combination of *biodiversity* and *bio-integrity* could provide quantifiable assessments of changes due to human impact at popular tourism venues, thus becoming a useful management tool.

Biodiversity was defined in Article 2 of the Biodiversity Convention, signed at the United Nations Conference on Environment and Development in 1992, as:

> the variability among living organisms from all sources including terrestrial, marine and other aquatic ecosystems and complexes of which they are a part: this includes diversity within species, between species and ecosystems.

Biological integrity or bio-integrity was defined over a decade earlier by Karr and Dudley as:

> the ability to support and maintain a balanced, integrated, adaptive community of organisms having a special composition, diversity and functional organization. (Karr & Dudley, 1981: 56)

Originally defined in the context of assessing water quality, bio-integrity failed to reach the wider audience that biodiversity gained from a UN Convention. 'Maintaining biodiversity' has become almost synonymous with ecological conservation. However, the two terms are complementary and powerful in tandem.

In its simplest form, biodiversity requires a comprehensive catalogue of species present in an area, preferably with indications of their population size, accompanied by periodic monitoring to ensure that all species remain present. It could, for example, be used to assess whether breeding birds, mammals or insects retain their status at heavily visited sites.

Bio-integrity shares the basic requirements of biodiversity, but also needs a closer understanding of dynamic relationships between the species and their environments. Indeed Crosbie (1998: 10) argues that 'bio-integrity', in the context championed by Minbashian, might be replaced by 'ecological integrity' or 'eco-integrity', defined in similar terms but implying the ability to sustain an integrated, adaptive *ecosystem* rather than just the system's community of organisms. This would take into account environmental as well as purely biological disturbances, for example the creation of footpaths due to the trampling of many feet, and the changes in drainage patterns and erosion that might ensue.

Both concepts acknowledge that the ecosytems to which they apply are subject to change. Both require original assessments followed by periodic monitoring, to be undertaken by trained personnel who know what they are doing and why. Should loss of diversity or shifts in integrity be detected, assumptions that the change is due solely or even partly to human interference may be quite wrong, even where human inroads have recently started or increased. Ecosystems are always changing, whether man interferes or not: as well as cataloguing and monitoring, both biodiversity and bio-integrity need careful interpretation against the broader background of whatever else is changing in the world. This point is discussed further below.

However, maintaining either or both of these forms of integrity are clearly objectives for good ecotourism. How they are achieved and monitored is discussed below.

Practicalities of Management

Arctic nations possess all the requirements to manage wilderness within their sovereign domains, and can if necessary agree among themselves to collectively manage trans-boundary wilderness areas. All can generate funding to realise management goals and objectives. In the Arctic, there is at least the potential for designating specific, allowable uses for wilderness, and a basic infrastructure for maintaining regulatory frameworks.

Best-practice management

Several internationally recognised best management practices are routinely employed to conserve environmental and heritage resources within wilderness regions. Widely accepted and simple, they represent the basic functions needed for competent management. Their components are as follows:

- *Creating a comprehensive inventory.* Resource management in wilderness areas requires knowledge of all existing resources, their condition, abundance and vulnerability to damage. Given the vastness of polar wilderness, difficult access and short seasons when fieldwork is possible, comprehensive inventories are time consuming and expensive. Useful knowledge of their dynamic ecological systems requires even longer-term investigations and higher costs. Whatever compromises are practicable, it is essential to have this baseline information that only field research can provide. Establishing benchmarks, designing effective monitoring programmes and application of management policies and plans all depend on the information revealed by an initial comprehensive inventory. A second inventory is also needed to identify and evaluate the social and economic uses of wilderness areas that various stakeholders are seeking – again difficult to obtain and organise, and often contentious (Wright *et al.*, 2001).

- *Identifying goals and objectives.* These are the mainstays of good management. In the case of wilderness management, enabling legislation usually defines government policies on intended uses. All too often it designates neither specific intentions nor management goals, instead expressing such laudable intentions as 'the long-term conservation of resources', avoiding specific objectives. This vagueness usually arises from lack of a satisfactory inventory (see above) or hard thinking that might identify more definite goals, but might also precipitate unwanted controversy. By default, goals and objectives are inferred and subject to interpretation, seriously reducing the effectiveness of the management planning process.
- *Creating management plans.* In the Arctic as elsewhere, the primary wilderness management objectives of tourism are to protect from harm both the resources being used and the tourists that are experiencing them. The management questions facing Arctic governments are when, where and how people may use wilderness without destroying it. Elsewhere throughout the world, government management agencies dealing with wilderness have responded in various ways (Loomis, 1993), ranging from strict restrictions of public access to participatory approaches defining allowable uses. The most severe restrictions seek preservation by preventing human entry. More collaborative approaches identify allowable uses, and then test their effectiveness by careful resource monitoring (Lucas, 1985; Stankey *et al.*, 1985). The prime importance of management goals and objectives in wilderness plans is in identifying allowable and prohibited uses – the foundations on which management programmes are implemented. Whether or not those uses are achieved is a measurable indication of success or failure of the planning.
- *Monitoring.* 'Monitoring' is a term often found in tourism management documents, all too often employed as a technical-sounding substitute for 'noting and recording'. In environmental management it has a more specific meaning: it is a method of detecting and measuring changes by collecting time-series of data for defined purposes, and observing trends in selected variables (Walton *et al.*, 2001: 33). Vital to all aspects of conservation management, it is used

particularly to identify alterations to ecosystems caused by human activity, beyond those caused by natural variations (Bertram & Stonehouse, 2007: 300).

How these components may be combined in multiple resource management planning is discussed in Chapter 9.

Best-practice management places particular emphasis on monitoring. Responsible ecologists cannot monitor without knowing what they are monitoring for, preferably within a management framework, and with the specific purpose of testing whether or not particular management objectives are being met. The ability to assess and manage tourism impacts in polar regions depends on an understanding of those relationships (Mieczkowski, 1995; Cater, 1994). Spellerberg (1991: 186), an experienced polar ecologist, stresses that simplicity and stability are essential for both the selection of parameters to be monitored and for long-term data collection. He summarises the value of monitoring as a means of:

- Establishing whether ecosystems and populations are being managed and conserved effectively.
- Assessing the best use of land.
- Indicating the state of the environment.
- Advancing knowledge of the dynamics of the ecosystem involved.

Figure 5.2 illustrates steps that in Spellberg's view need to be taken to establish monitoring for a management programme, that would check the progress of the programme at several stages, and ultimately show whether selected measures are effective or not.

Monitoring to ecological standards is expensive and time consuming. Quasi-monitoring – observing and recording without clear objectives, undertaken to satisfy a legal requirement for 'monitoring' – invites derision. There are many stages between, which provide at least some of the required information without vast expenditure.

Implementation in backcountry areas

Much of the Arctic is now governed, at least in part, by customs and traditional laws established and enforced by native communities. The Nunavut in Canada, the Inuit in Greenland, the Saami in Scandinavia and native peoples of Russia and Alaska are determining allowable recreational uses of their land and water. Terms and conditions expressed in their management approaches offer valuable examples that can be applied throughout the polar regions.

However, politics and competition for budgets are perpetual management issues affecting the governance of Arctic tourism. Officials responsible for wilderness recreation management realise that success

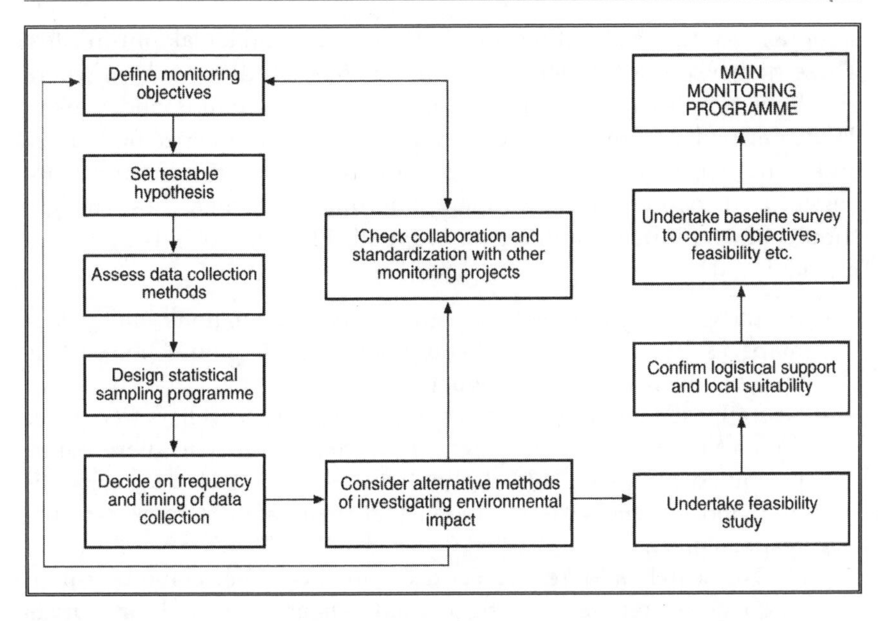

Figure 5.2 Steps to be used in the design of an environmental monitoring programme. (Adapted from Spellerberg 1981: 182)

in the political and budget arenas depends on providing quality tourism experiences to the public and economic benefits to local communities. They must demonstrate that the fees they collect from recreation activities and associated economic benefits to local communities justify their budgetary requests. Recreation managers have navigated this difficult course for a long time, and make considerable efforts to demonstrate that recreation participation is strong and growing.

Absence of development may sustain the integrity of wilderness, but is an obstacle to environmental and tourism management. The Arctic's lack of transportation systems, telecommunication networks, trained personnel and service facilities impose management constraints. Operational functions essential for meeting the demands of large numbers of tourists, such as resource conservation and monitoring, scientific research, security patrolling, visitor safety, waste collection and disposal, and emergency response capabilities are all weakened by scarce infrastructure.

Backcountry resource managers adapt to these constraints by restricting numbers of tourists allowed in particular areas, and ensuring that, where waste collection or emergency services do not exist, tour groups are self-reliant. By widespread publicity, they designate specific entry points or gateways to intercept tourist before they start, and a network of trailheads, information kiosks, interpretive centres, ranger stations, boat ramps, hunting zones, fishing sites and kayak put-in sites. These dispense information and simultaneously indicate when, where and how people are travelling through wilderness lands and waters. Thus, particular forms of recreation are encouraged in specified areas, and tourists provide information on how and when the resources are being used. Monitoring can then determine whether these usages should be continued, modified or stopped. Elements of this approach include:

- At the gateways, advice to sign in, read regulations and obtain maps, essential supplies and equipment including guidebooks. This may be given in several languages.
- At trailheads and boat-launch sites, requests to travellers to register their route and mode of travel. They register also at intermediate permitted campsites and destinations. Known hazards are identified, and travellers are reminded that their safety is their personal responsibility.
- Where search and rescue services are available, communication frequencies and weather radio broadcasts are provided for tourists with communication equipment. Boat and kayak travellers are required to possess vessel licenses and have marine charts, tide tables and adequate safety and navigational equipment on board. Coast guard vessels are rarely available to conduct inspections.
- Rangers patrol from fixed or seasonal bases by boat, aircraft and various modes of land travel, though limited by ruggedness of the terrain, weather conditions and budget considerations.
- Search and rescue and emergency response capabilities vary widely with area, remoteness, weather conditions and budgets. Remote Arctic settlements rely mainly on volunteers for this service.
- Resource monitoring and evaluation is conducted by rangers and others, at frequencies that depend on the budgetary and personnel resources. Lead agencies may be helped by NGOs that commit volunteers, raise funds and public awareness, and advocate policies, providing invaluable contributions to conservation of Arctic resources. Tourists are encouraged to actively support these organisations.

Such management techniques are widely used in national parks, wildlife refuges, forest reserves and other sites (including UNESCO World Heritage and Biosphere Sites). Site management practices, participatory decision-making processes, partnerships with stakeholders and commitment to international resource conservation agreements are all embodied in the rôles and responsibilities of the resource agencies involved.

Licensed guides and special-use permits

A successful management technique for conserving Arctic resources and promoting lawful visitor behaviour is guide licensing. Wildlife managers realised long ago that an effective way to ensure regulatory compliance was to require anglers and hunters to employ licensed guides. Guide licensing has been established by management agencies in all Arctic nations. Guides are instructed in specialised knowledge of environmental conditions, resource laws and regulations, and survival, emergency response skills and facility in observing and reporting. In most jurisdictions, refresher courses are required to sustain both educational knowledge and practical skills. Licensing in most jurisdictions has extended beyond angling and hunting to mountaineering, rafting, kayaking and wildlife photography.

Guides so appointed evolve from apprenticeship to master status. They are held directly responsible for both resource protection and visitor safety. License requirements include insurance coverage, bonding and indemnification. Based on the risks they are legally required to accept, guides diligently monitor the behaviour of their clients. They accept the professional and financial risks of their profession for the privilege of pursuing a unique way of life, generally spend a considerable time teaching their clients appropriate behaviour and respond quickly when clients willfully disregard their information. Licensing benefits tourist behaviour, and can be used beneficially to restrict numbers by requiring that visitor groups are accompanied by guides.

A further way of restricting numbers is the issue of special use permits to native people for conducting river rafting, mountaineering and wildlife viewing expeditions. Native people and local residents who depend on polar resources for their cultural and economic well-being are the strongest advocates of sustainable polar resources, and as guides they tend to provide wise stewardship.

Antarctica has no such guiding or guide-licensing programmes, though many who are concerned with current tourist guiding would welcome something of the kind. Currently, Antarctic tourism 'guides' on the best-run ships are knowledgeable persons employed by the cruise operators to lecture on board. Often with distinguished records of scientific research and a thorough knowledge of the area, they may lack any formal training in the skills and techniques that are mandatory in Arctic guiding. On less reputable operations with smaller budgets, 'guides' may include well-intentioned volunteers from the hotel staff, who do little more than help passengers ashore and stand by to take photographs.

As numbers of Antarctic tourists increase, more competent guides will be needed. It is not difficult to see that proficiency in skills, including subject expertise, knowledge of landing sites, emergency response and communication skills, and tourist management, should be included in their job specification – skills that will help to protect both Antarctic resources and the safety and well-being of the tourists.

Summary and Conclusions

This chapter introduces the concept of wilderness, and polar regions as the world's greatest wilderness areas that require special management to maintain their wilderness qualities. Though the Arctic has been exploited for mineral resources over three centuries, and exploitation may intensify rather than diminish, those responsible for its management are obliged to maintain its wilderness qualities as best they can. Every environment that is subject to tourism requires management organisations with clear objectives (provided by tour operators), operating within regulatory frameworks (provided by governing authorities).

To be effective, both operators and authorities need to use terms to which clear meanings can be attached, and to make sure that they agree on the meanings ascribed. This prescription is especially important in polar regions, which may be particularly vulnerable to human incursions. However, it is especially likely to be evaded because these are difficult and expensive regions for both sides – for operators to conduct their business, or regulators to know and provide appropriate regulation.

We examine criteria for successful management, in particular 'sustainable' and 'responsible' tourism, 'ecotourism' and other systems that attempt to set standards of environmental protection. Does 'ecotourism' meet the case for the special standards of tourism practice required in polar regions: are 'biodiversity' or 'bio-integrity' the important qualities to maintain? Finally, we discuss the realities of how tourist management operates in the Arctic, but cannot operate in the Antarctic for lack of firm objectives and revenues.

Chapter 6
Managing Shipborne Tourism

Introduction

Shipborne tourism, the largest and fastest-growing segment of polar tourism, is dominant throughout both polar regions. It appeals to millions of passengers aboard cruise and expedition ships, to the smaller numbers in chartered ships seeking wildlife viewing and sport, and to individual adventure mariners in vessels ranging in size from motor yachts to kayaks. Two factors contribute to the certainty of its future growth:

- Polar tourists constantly seek more diverse destinations, recreational activities and shore excursion experiences, all objectives that the cruise ship industry uniquely sustains.
- Arctic governments encourage expansion by increasing facilities for more and larger vessels, providing facilities at new destinations and prolonging tourist seasons. For the Antarctic, encouragement comes not from the ruling Antarctic Treaty System, but from gateway ports that benefit by servicing the industry.

Growth of polar shipborne tourism is consistent with the growth of the industry worldwide. According to the Cruise Lines International Association, the number of passengers on cruises throughout the world has grown from about 500,000 in 1970 to more than 12 million in 2006. New and larger cruise ships constantly enter the market. From 2000 to the end of 2008, 88 new cruise ships were added, including the Royal Caribbean's *Freedom of the Seas*, which entered the fleet in June 2007 with a passenger capacity of 3634 – twice that of big ships built only a decade ago. In the Arctic as elsewhere, the industry's growth is supported by governments' promotional and infrastructure investments, as well as improved transport technologies, increasing wealth and leisure time, and worldwide climatic moderation (Brown, 2006).

In the Arctic region, at least 80 cruise ships were booked to dock in Reykjavík harbour in 2009, bringing between 55,000 and 60,000 tourists ashore. Despite the economic crisis in both Iceland and the rest of the world, tourists keep flocking to this destination (*Iceland Review*, 1/3/08). The number of tourists coming to Northwest Russia in 2008 exceeded 12 million people. Experts say this number could increase to 17–24 million by 2010 (*Barents Observer*, 3/11/08). The 2008 season has witnessed the start of cruise ship operations through the Northern Sea Route and in the Kara, White and Barents seas – all operations conducted aboard Russian icebreakers. During the past 20 years, cruise ship companies have greatly expanded their passenger capacity and itineraries, while simultaneously new entrants have diversified the cruise ship market. In the 2008 season, for example, a single cruise ship company offered 57 distinctly different Arctic tours aboard a fleet of 18 vessels.

Shipborne tourism to the Antarctic region has increased on a more modest scale, but still considerably and at an accelerating rate: according to Table 3.1 (p. 50) and more recent IAATO data:

- in 1979–1980, two ships made a total of four voyages, carrying 855 passengers;
- in 2000–2001, 18 ships made 116 voyages, carrying almost 12,000 passengers;
- in 2007–2008, 55 ships made 291 voyages, carrying over 45,700 passengers.

Numbers of passengers is a vitally important statistic, but not the only one to consider for the purpose of responsible tourism management. The crew, staff, shipboard lecturers and guides, often numbering in the hundreds on large vessels, must also be included in emergency response plans. Comprehensive emergency response planning and shore-based infrastructure capacities must account for the total number of humans at risk. When viewed from this perspective, the true dimensions of the dangers associated with tourism marine incidents are readily apparent.

Growth in either polar region presents significant management challenges to operators, governments and local communities. Even the best-run ships are arguably more at risk in polar waters than on the better-known and more popular temperate and tropical routes. Severe maritime conditions, lack of accurate navigational information, shortage of emergency response infrastructure and unreliable communication networks are some of the special hazards involved. This chapter records major and near-catastrophic marine polar tourism incidents that have occurred, the legislation and guidelines within which polar ships operate, and the management policies, plans and physical resources needed for effective management.

Polar Marine Incidents

Numerous types of marine incidents occur in polar waters that incapacitate ships and endanger both passengers and crews. In terms of threats to human life, these incidents range in severity from sinkings to the loss of propulsion. In addition to the actual or potential harm to humans, marine incidents also involve damage to the environment including oceans, the atmosphere and wildlife resources. The actual numbers of polar cruise ship incidents are presented in Table 6.1 and

Table 6.1 Marine accidents involving polar cruise ships 1979–2009

Marine incident	*Total events*	*Events since 2000*	*Percentage since 2000*
Polar cruise ships sunk (1979–2009)	8	5	63
Polar cruise ships running aground (1972–2009)	27	16	59
Pollution and environmental violations (1992–2009)	64	42	65
Disabling by collisions, fires, propulsion loss (1979–2009)	34	28	82

Source: www.cruisejunkie.com

these include sinkings, groundings, pollution and environmental violations, and disabling by collisions, fire and propulsion loss.

One consequence of the growing popularity of cruising in polar waters is a significant increase in marine accidents that have occurred since 2000 (Table 6.1). As Table 6.1 shows, in the period 2000–2009, five polar cruise ships sank, 16 were grounded and 28 were disabled by collisions, fires or propulsion loss. Another 42 polar cruise ships committed environmental damage. All those incidents placed at risk several thousand passengers, crew and staff. Incidents included:

- *Explorer*, an ice-strengthened cruise ship that sank in Bransfield Strait, off Antarctic Peninsula in 2007. The 154 passengers and crew were fortunate in having calm weather and another cruise ship close by at the time.
- *Disko II* in the same year ran aground off Greenland's west coast. Again, good weather allowed another cruise ship to take off its 52 passengers and two tour guides without incident, while the crew of 18 remained on board.
- *Clipper Odyssey* in 2006 ran aground in the Aleutian Islands, forcing 153 passengers and crew to transfer to other ships.
- *Mona Lisa* in 2003 ran onto rocks near Spitsbergen, in the Barents Sea: 670 passengers were evacuated.

In each of these accidents, a fortunate combination of calm weather and the proximity of another ship mitigated dangers to passengers and crew, no doubt helping to prevent loss of life. Neither calm weather nor company can be relied on, even at the height of summer. Such incidents are likely to increase as numbers of ships and voyages increase. The threats arising from them need to be understood in order to design and implement effective management responses.

It must be noted that polar marine incidents affecting tourists are not exclusively the province of cruise ships. Coastal ferries, charter touring vessels, sportfishing vessels, private yachts and kayaks complete the fleet of vessels transporting tourists through polar seas. All these types of vessels have experienced marine incidents that resulted in threats or loss of life as well as the dedication of enormous resources for search and rescue operations.

The management implications of increased tour vessel traffic are clear: more ships in polar waters imply increasing threats to human safety and environmental integrity. They will require increasingly efficient management responses, notably, quick-response search and rescue capability to save human lives, investment in strategically placed infrastructure, land-based accommodations to house, feed and medically attend to rescued passengers, and containment and restoration capability to prevent or minimise environmental damage, notably oil spills.

A detailed discussion of specific management actions required to enhance safe ship passage in polar waters was produced by the Arctic Council (2009) in their report entitled 'Arctic Marine Shipping Assessment 2009 Report'. The report indicates that increasing the safety of polar cruise transits and responding to emergencies of all sorts associated with those cruises is based on two principles. The first is to prevent harm, the second to improve incident response capabilities. A summary of specific actions needed to accomplish those principles is provided in Box 6.1 (p. 105).

Sources of Protective Legislation

Resources for safe navigation, regulation and management of marine operations are generally found between 60° north and south latitude, where nearly all the world's shipping lanes are located. Management for seas beyond 60° is more difficult to achieve for reasons already discussed. However, international treaty conventions and national laws governing marine safety and environmental protection extend into polar regions, and polar shipborne tourism has adopted regulations and guidelines provided both by the industry itself and by interested non-governmental organisations. Laws and regulations are enforceable where enforcement is available. Guidelines are not, but usually represent best practices and are therefore acceptable by a majority of responsible operators.

International conventions and regulations

International conventions and agreements for protecting human life and preventing pollution at sea are in force worldwide. Table 6.2

Box 6.1 Safe passage in polar waters

Actions to prevent harm
Information for safe passage

- Ice Condition Information: Expand Coverage
- Weather Information: Expand Coverage and Timely Notifications
- Update Hydrographic Charts
- Navigational Aids
- LRIT Long Range Identification and Tracking
- Coordinate Cruise Ship Transits & Scheduling for Mutual Aid
- Collaboration of International Information Organizations (e.g. Maritime, Hydrographic, Meteorological, and Maritime Satellite Organizations)

Guidelines for vessel operations

- IMO Guidelines for Ships Operating in Arctic Ice-covered Waters
- Improve Passenger Ship Safety in Arctic Waters
- Multilateral Arctic Search and Rescue (SAR) Instrument
- Unified Governance: UN Convention on Law of the Sea

Actions to improve incident response capabilities
Expand infrastructure capacity

- Places of Refuge
- Expand and Develop Ports
- Search and Rescue Resource Investments
- Medical Evacuation & Care Resources
- Shoreside Evacuation Shelters and Provisions
- Environmental Incidence Response Equipment and Personnel
- Salvage Resources
- Waste Disposal Facilities
- Law Enforcement Resources

Human resource training

- Mariners
- Ice Navigators
- Emergency Service Providers
- Lifeboat Drills and Prepare Tourists for Extreme Conditions
- Environmental Managers and Monitors

lists in chronological order some of the more important laws, regulations and guidelines that apply to polar marine operations. More details of their provisions appear in Appendix C. Most of these enactments concern prevention of pollution, suggesting that threats to

Table 6.2 International conventions and agreements for protecting human life and avoiding pollution at sea

1969	Convention Relating to Intervention on the High Seas in Cases of Oil Pollution Casualties
1970	Arctic Waters Pollution Prevention Act
1972	Convention on Prevention of Marine Pollution by Dumping of Wastes and Other Matter (London Convention)
1973	International Convention for the Prevention of Pollution from Ships as modified by the 1978 Protocol (MARPOL 73/78)
1974	International Convention for the Protection of Life at Sea (SOLAS)
1982	United Nations Convention on the Law of the Sea (UNCLOS)
1990	Oil Pollution Preparedness, Response and Cooperation Convention (OPRC)
1991	Arctic Environmental Protection Strategy (AEPS)
1995	Washington Declaration and Global Programme of Action for the Protection of the Marine Environment from Land-Based Activities
2002	International Code of Safety for Ships in Polar Waters (Polar Code)

Note: For further details of each see Appendix C

the environment are currently the major concern. However, the first consideration in any emergency at sea is the protection of human life, with which the 1974 International Convention for the Protection of Life at Sea (SOLAS) deals adequately. The date is misleading: originating in 1914, SOLAS is one of the earliest international agreements, and its provisions, like those of the other conventions and agreements, are under constant review.

Such international agreements provide the safety legislation and regulations within which the sea-going ships of all signatory states operate. This includes virtually all polar cruise ships, which may be subject to inspection and certification to ensure that they conform to requirements. Day-to-day activities of passengers, staff and crews aboard the ships depend more on legislation by local authorities, and on guidelines agreed between cruise operators and other interested parties.

National laws and regulations

The eight Arctic nations have enacted, and currently enforce, laws and regulations governing marine operations and pollution prevention within the boundaries of their territorial waters, providing a framework of cooperation to promote human safety, protect the environment and provide coordinated responses to marine emergencies. They are now dealing specifically with issues, including problems of growing congestion, arising from cruise ship operations.

Arctic governments find themselves in the dilemma of simultaneously trying to prevent personal and environmental harm while aggressively promoting tourism. Ideally, the application of management practices that enhance marine safety and prevent environmental damage will result in improved environmental stewardship and reduced risk to tourists. But concerns about protecting jobs, economies and communities also prevail throughout the Arctic's circumpolar world.

National attempts to regulate marine tourism extend from exceedingly stringent controls to more flexible management techniques. The Norwegian government, for example, plans to restrict cruise ship traffic around the archipelago of Svalbard, and prohibit the use of heavy fuel oil. In a public announcement of this policy, Norway's Minister of the Environment, Helen Bjørnøy, stated that it was important to limit the number of tourists visiting nature preserves in the Svalbard area: 'the goal is to hinder spills that could have hugely negative consequences for the environment in the fragile and valuable areas around Svalbard' (*Aftenposten*, 6/4/07). The new rules will limit to 200 the number of passengers allowed on board each ship that enters nature preserves on East Svalbard and those tourists who are allowed entry are paying a special environmental tax.

Another approach to the management of polar marine tourism, instituted by the US State Department and the State of Alaska in 2008, is the Alaska Cruise Ship Initiative. This provides for rangers (who are marine engineers trained and licensed by the US Coast Guard) to travel aboard large cruise ships, to monitor wastewater discharges in Alaskan waters and compliance with other pollution requirements for the state (*Juneau Empire*, 6/7/07; *World Maritime News*, 2/27/08). The programme will cost each passenger $4.

Concerns about managing marine tourism are major issues for remote Arctic communities. The residents of Kodiak, Alaska, by example are now attempting to determine the 'ideal number of cruise ships' they should host. For a decade they hosted eight or nine a year, but in 2008 the number had doubled to 16 and in 2009 the number of scheduled arrivals increases to 24. Kodiak is representative of the attempts of Arctic communities to balance economic benefits with infrastructure costs, environmental protection and cultural intrusion (*Kodiak Daily Mirror*, 10/23/08). Their remote locations, and limited financial and personnel resources severely constrain effective management.

Threats to Human Safety

As shipborne tourism grows, more tourists are at risk, and their safety is the dominant consideration in all management issues facing governments, communities and the industry. However, the large number of cruise ships currently in polar waters already exceeds emergency response capabilities that local authorities can provide (US National Ice Center, 2007).

Cold air and water require quick and efficient rescue: even limited exposure quickly reduces chances of survival. Table 6.3 indicates that a subject inadequately dressed for immersion in cold water is likely to die from drowning within an hour. The point of incapacity would be reached sooner in water closer to freezing temperature. In polar search and rescue, time is indeed the enemy.

As numbers of cruise ship passengers and crews rise, numbers of medical emergencies – from life-threatening heart attacks to sprained limbs – cannot fail to increase in proportion. Occasionally, bizarre accidents occur, such as the 2007 incident along the Svalbard coast where 18 Britons standing near the railing of their vessel suffered injuries (three serious) when blocks of ice from a glacier fell on them (McGrath, 2007; *Aftenposten*, 8/9/07). Minor incidents may require no more than stabilisation of the patients. More serious incidents, which are beyond the capacity of ships' medical services, require evacuation.

During the time that tourism has become the single largest human presence in polar regions, demands for sophisticated medical services, including evacuation from remote locations, have increased substantially, placing severe burdens on infrastructure. In both polar regions, networks of medical advice and support are now available to medical staff at sea through international radio channels, polar stations and other ships. One positive advantage arising from the increasing numbers of ships involved is the increased range of medical facilities that they bring to

Table 6.3 Survival times of lean subjects in rough seas at 6°C (42.8°F)

Clothing type	*Hours to incapacity*	*Hours to unconsciousness*	*Hours to cardiac arrest*
Light clothing	0.4–1.3	0.8–2.6	1.3–4.3
Wet suit (4.8 mm)	1.6–4.7	3.1–9.9	4.9–16.2
Insulated dry coveralls	2.9–8.8	5.7–18.2	9.1–30.0

Source: Data from Sullivan (1998: 84)

Note: Incapacity, induced at a body core temperature of 34°C (93.2°F), is likely to result in drowning before further cooling induces unconsciousness (30°C, 86°F) or cardiac arrest (25°C, 77°F)

the area. For further discussion of medical services available to passengers on cruise ships, see Levinson and Ger (1998).

Cruise ships are particularly prone to outbreaks of gastrointestinal infections, due to noro virus and other micro-organisms. In 2002–2007, 34 such outbreaks worldwide were reported to the Center for Disease Control, involving over 4000 passengers and crew. In 2007 alone, seven polar cruise ships were affected. Though such incidents cause inconvenience rather than danger, and can usually be dealt with on board, Arctic health officials may at times need to respond to severe outbreaks of shipborne infections. The major cities of the eight circumpolar nations have excellent health care infrastructure, but small high Arctic communities would be unable to respond to illnesses involving hundreds of cruise ship passengers and crew. In view of their isolation, they would themselves be particularly endangered by contact with infectious diseases from outside.

A less recognised threat to polar tourists is exposure to a variety of dangerous contaminants in structures and relics of the Cold War, historic settlements, mines, canneries and scientific stations. These were usually abandoned without environmental remediation, and are now in a dangerous condition due to weathering. In addition, they may harbour quantities of asbestos, lead-based paints, scrap metal, munitions, diesel fuel, PCBs, pesticides and heavy metals, many of which have seeped unchecked into soils and rivers (*SIKU News*, 8/27/07). Alaska is currently attempting to clean up the many old barracks and radar stations located along its entire coast and on the Aleutian Islands. All will require extensive and very costly remediation. Until this is completed, tourists should regard them as areas offering serious hazards.

The Soviet Union's environmental legacy similarly includes many contaminated areas and sites. Radioactive substances, heavy metals and other untreated wastes were disposed of in Arctic rivers and seas, and the Arctic landscape generally has been severely abused with toxic substances. The Russian Federation is attempting remediation, a task on an enormous scale, but necessary before the tourist potential of its Arctic region can be fully realised.

Threats to the Environment

Polar marine tourism can cause environmental harm in numerous ways, from sinking and grounding of ships at sea to inappropriate behaviour of passengers ashore, requiring international conventions and national laws and regulations to minimise opportunities for damage; see also Vidas (2000).

Marine tourism operators, prompted by legal mandates, chastised by large punitive judgements and motivated by self-interest to protect their reputations in a highly competitive market, try not only to comply with environmental statutes, but implement self-imposed guidelines for environmental protection. To date, cruise ship operators, industry associations, governments and non-governmental organisations have successfully collaborated to minimise environmental impacts. Their efforts, both singly and collectively, have created several best-management practices. However, the growth of Arctic marine tourism will require a substantial increase in effort in the future. This section distinguishes the several types of ship-related incidents that can impact polar environments, identifying the management challenges that face responsible parties in both private and government sectors.

Oil spills

Between 1993 and 2009, cruise ships on polar voyages were involved in more than 40 fuel spill incidents. Some were the result of such serious accidents as sinking, grounding and collision. The cruise ship, *Explorer,* which sank in Bransfield Strait, Antarctica, in November 2007, had 185,000 litres of fuel oil, 1000 litres of gasoline and 24,000 litres of lubricants on board, leaving an oil slick 5 km wide by 8 km long (Reel, 2007).

The environmental damage resulting from marine oil spills is all too well known. Spills in any ocean are difficult to contain because they are dispersed erratically by currents, waves and winds. In polar regions, they are particularly difficult to manage because sea ice mixes with the oil and interferes with containment booms. The oil products destroy all aspects of the environmental integrity of the marine ecosystems, including fisheries, marine mammals, corals, ocean and shore birds, and both pelagic and coastal wildlife.

Timely and effective response requires containment of the spill, recovery and restoration. Remoteness of polar venues from normal shipping lanes may delay the start of containment and recovery operations. Strong winds, rough seas, ice and working in cold temperatures all contribute to the difficulties. The circumpolar Arctic nations perceive the need for innovative approaches to the problems of oil spills at low temperatures. Research to identify effective technologies and management techniques is being conducted at the Coastal Response Research Centre, University of New Hampshire.

Regrettably, fuel spill incidents have also resulted from blatant violation of law. In July 1999, one prominent company pleaded guilty in six jurisdictions to charges of fleetwide practices of discharging oil wastes and other effluents, including into the waters of Alaska's Inside Passage. On all charges the company was fined $18 million dollars (General Accounting Office, 2000, *et. seq.*)

Sewage, wastewater and bilge disposal

In 2004, the US Commission on Ocean Policy reported that, on average, a cruise ship generates 140,000–210,000 gallons of sewage and a million gallons of wastewater from sinks, showers and laundries each week (APRN, 12/31/07; Choi, 2007). Similarly, the Environmental Protection Agency found that the average cruise ship passenger generates about eight gallons of sewage a day while at sea (APRN, 12/31/07; Choi, 2007). Sewage carries germs that can contaminate shellfish beds and harm other life, while phosphates, nitrates and other wastewater compounds can trigger huge growths of algae that cloud the water, reduce oxygen and kill fish.

In 2001, the 21 members of the Cruise Lines International Association agreed not to release wastewater within four miles of shore and took action to prevent this type of environmental pollution. By 2007, the Association reported that approximately 40% of its members' 130 ships, which make up two thirds of the world fleet, had installed advanced wastewater systems, with 10 or 15 more added every year. Each system costs $2 million to $10 million per ship and takes six to twelve months to install (Choi, 2007).

Despite these and similar efforts, laws created to control wastewater discharge are still violated in polar regions as elsewhere. The problem of wastewater disposal is especially acute in the Arctic, where shore-based disposal and treatment infrastructure is extremely limited. Between 2000 and 2007, six violations for the unlawful discharge by polar cruise ships resulted in court judgements against companies and the levying of fines (General Accounting Office, 2000, *et. seq.*). In addition to fines, there are the real but intangible costs of tarnished reputations. The companies involved have not only paid restitution, but have also contributed substantial sums of money to environmental conservation organisations and projects.

A cruise ship may also produce more than 25,000 gallons of oily bilge water from engines and machinery a week, according to a 2000 Environmental Protection Agency report (Choi, 2007). Numerous international conventions regulate the handling and disposal of oil in the marine environment, with special provisions for the disposal of bilge. All the Arctic nations are signatories to the conventions and seek to enforce them; cruise ship companies are fully aware of the need for compliance. The difficulty faced by governments and companies is the lack of infrastructure to appropriately dispose of bilge or respond to a spill incident.

Solid waste disposal

Establishing acceptable methods for solid waste disposal is a big challenge in the Arctic. There is agreement among marine operators and governments that the best solution is to stow debris on board for the duration of the cruise. However, when tourist vessels arrive at Arctic ports, the capacities of land-based disposal sites are not sufficient to dispose of solid waste. Onboard incineration raises concerns about air pollution, with implications that emissions may release dioxins and fine particles that can cause respiratory ailments. In select areas where offshore disposal of certain degradable items is allowed, there is concern that ice and water temperature conditions will delay or prevent the degradation process.

The difficulties associated with solid waste disposal in the Arctic are significant and unique. Traditional solutions, such as landfill sites, are prohibitive because of environmental conditions such as permafrost, either solid or thawing, lack of soil, snow and ice, and wildlife scavenging. Compaction technologies are effective, but costly in terms of equipment, trained personnel and transport. Incineration is appropriate if extremely high temperatures can be sustained, but prevailing dampness and the cost of energy can reduce the efficiency of this approach. Containerised exportation is seemingly ideal, but very expensive. For those communities that have regular barge freight service, containerised export is more feasible.

For Antarctic cruises, the practice has arisen of separating and stowing solid waste aboard, and disposing of it against payment to the municipal authorities of the nearest gateway port. Few voyages last more than 10–14 days between ports, so the accumulation is never great. How the port disposes of it is a matter for its own authorities to decide: in poor or growing communities, payments from cruise ships may help to provide better facilities for waste disposal generally.

Air pollution

The naturally clear conditions of the unsullied Arctic atmosphere can provide brilliant displays of ice, sunsets, terrain and aurora borealis. However, visibility may be impaired by fogs, some of which are generated by air pollution, to which Arctic communities are not unreasonably sensitive. In any Arctic port, an obvious source of pollution is a ship's smokestack, particular on firing-up before departure. The need to reduce such emissions is a concern of the International Maritime Organisation (IMO), which in 1997 adopted regulations to limit sulphur dioxide and nitrogen oxide emissions from ship exhausts. Virtually all cruise lines operating in polar regions are now implementing new technologies, using different fuels and employing new operational practices to comply with IMO regulations.

Some technologies are experimental, with debatable effects. A demonstration project using sea water to scrub sulphur and other contaminants from engine emissions received the cooperation and backing of several Canadian and US government and regulatory agencies. It was subsequently rejected when Swedish and British research showed that sulphuric acid produced in the process, and washed back into the sea, freed carbon dioxide from surface waters (Province, 2007). Other technology innovations have been more successful. Celebrity Cruises, for example, has four ships with gas-turbine engines, which release 80–95% less sulphur, fine particles and nitrogen oxides than marine diesel engines.

Between 2000 and 2007, several convictions and fines were brought against cruise ship operators who caused pollution in Arctic National Park and Marine Sanctuary waters (General Accounting Office, 2000, *et. seq.*). Those penalties sent a clear message to the industry, which has resulted in more appropriate ship operations. However, air quality monitoring stations and enforcement resources are equally rare in polar environments, and the cumulative effects of several ships operating in relatively constricted zones have yet to be assessed.

Threats to wildlife

Wildlife is a primary attraction for tourists at both ends of the world. Polar ecosystems, particularly in coastal environments, provide opportunities to view many species of land and marine mammals and a remarkable diversity of birds. Cruise and expedition ships seek locations where those attractions can be seen, though with an inherent risk of disturbance. Recognising the need to protect both wildlife and the safety of their tourists, operators employ wildlife viewing guidelines (Figure 6.1). Both the International Association of Antarctica Tour Operators (IAATO) and the Arctic Expedition Cruise Operators (AECO) have created self-imposed guidelines for both ship operations and the behaviour of their passengers when ashore. These have been remarkably successful, as acknowledged by the Antarctic Treaty System in the south, Arctic governments in the north and such non-governmental organisations as the World Wildlife Fund (WWF) and Conservation International. For details of these initiatives, see **'Infrastructure and information'** (p. 116).

Cruise ships are inherently large, noisy and intrusive, operating in a medium that transmits sound over long distances. Appropriate operations among wildlife include:

- Avoiding disturbance of whales and other marine mammals at sea. Constant vigilance by navigating officers reduces possibilities of such general disturbance as separating mothers and calves. Moving at slow speeds reduces risks of collisions. Keeping clear of known feeding areas reduces turbulence and interference with a vital activity. Slow speeds reduce propeller noise that may interfere with

Figure 6.1 Wildlife encounters by cruise ship must avoid harm to both people and animals. Photo: JMS.

the animals' sonar. On meeting whales, experienced captains back off and wait for the animals to approach them (*SIKU News*, 4/4/07).

- Avoiding disturbance in critical habitats, for example in seal-breeding areas of pack ice where pups lie out on ice floes, or near bird colonies during breeding times when nestlings are easily subject to panic. Onshore encounters with walruses, fur seals or polar bears can prove dangerous anywhere and at any time.
- Avoiding habitat intrusion. When shore visits are allowed, tourists need thorough briefing before landing and careful supervision while ashore. Numbers especially need to be manageable, and behaviour observed by qualified guides. Ship operations involving shore excursions have to be well planned and coordinated with those of other ships.
- Avoiding pollution. Discharge of bilge and wastewater in unauthorised areas can interfere with local fish stocks and fisheries on which local residents may depend. Environmental damage to marine ecosystems from unlawful discharges is still enormous, and economic losses for commercial fisheries, sport angling and native people reliant on fisheries are extensive.

Guidelines for Good Practice

Expedition cruise ship companies operating in both polar regions have created guidelines to enhance marine operations and visitor safety, and provide environmental and cultural resource protection. The practice began with the formation of the International Association of Antarctica

Tour Operators (IAATO) in 1990, which as one of its first activities produced guidelines for ship operations in Antarctic waters and for the behaviour of visitors ashore. Later, the guidelines were enhanced to include emergency response plans, the protection of Antarctica's marine and land resources and the preservation of the southern continent's heritage resources (see below). Later initiatives by the WWF and the parallel association of AECO provided similar guidelines for Arctic cruising.

International Association of Antarctica Tour Operators

IAATO's Emergency Contingency Plan (www.IAATO.org.) has been successfully implemented on several occasions and is constantly updated to improve emergency response capabilities. IAATO's guidelines for the protection of visitors, environmental and heritage resources have been in effect for nearly two decades. Given the fact that these guidelines are directly relevant to polar conditions, marine tourism operations and the management of tourists when ashore, Arctic governments, communities and tour operators would most probably benefit from their application to Arctic tourism. IAATO guidelines for visitors are provided in Appendix D.

World Wildlife Fund Arctic programme

The WWF International Arctic Programme sees tourism as one way to support protection of the Arctic environment. Tourism can be conducted responsibly so that visitors learn to appreciate and respect Arctic nature and cultures, as well as provide additional income to local communities and traditional lifestyles (www.panda.org, 2007). In 1995, the Arctic Programme began developing principles and codes of conduct for Arctic tourism, and a mechanism for implementing them. The goal was to encourage the development of a type of tourism that protected the environment as much as possible, educated tourists about the Arctic's environment and peoples, respected the rights and cultures of Arctic residents and increased the share of tourism revenues that go to northern communities.

The Principles and Codes for Arctic Tourism were developed in cooperation between WWF Arctic Programme, tour operators, conservation organisations, managers, researchers and representatives from indigenous communities during workshops held in Svalbard in 1996 and 1997. The participants developed a *List of Potential Benefits and Potential Problems of Arctic Tourism*, *Ten Principles for Arctic Tourism*, a *Code of Conduct for Tour Operators* and a *Code of Conduct for Tourists*. The *Principles* and *Codes of Conduct* are provided in Appendix D.

Association of Arctic Expedition Cruise Operators

The AECO was founded in 2003 for the purpose of managing:

> respectable, environmentally-friendly and safe expeditions in the Arctic. The members agree that expedition cruises and tourism in the Arctic must be carried out with the utmost consideration for the vulnerable natural environment, local cultures and cultural remains, as well as the challenging safety hazards at sea and on land. AECO members are obligated to operate in accordance with national and international laws and regulations and agreed-upon AECO by-laws and guidelines.

AECO's offices are located in Longyearbyen, Svalbard, Norway (AECO.org.no). AECO developed its guidelines with considerable input from the Governor of Svalbard, the Norwegian Polar Institute, the WWF Arctic Program Office, Greenland Tourism, Greenland Directorate of Environment and Nature and other interested parties. AECO's geographic influence is limited to Svalbard, Jan Mayen and Greenland, but expansion into other Arctic regions is under consideration. A note on their guidelines appears in Appendix D.

Infrastructure and Information

Arctic nations, both individually and collectively, are legally responsible for providing infrastructure for the health, safety and welfare of citizens and visitors, including that which is needed to provide marine management support. Managing marine tourism requires coastal authorities to know where ships are, to respond to marine incidents, control traffic numbers, enforce laws and regulations and disseminate vital information. Efforts to improve their capabilities are now being implemented throughout the Arctic. Accurate information on ice and weather conditions is provided by ice centres and weather stations of the circumpolar nations, using satellite, photo imagery and telecommunication technologies. These facilities are constantly updated. Advanced ice centres, such as the Canadian Ice Service, can now precisely monitor spill incidents, a capability that greatly enhances environmental monitoring and facilitates faster incident response.

Arctic jurisdictional authorities possess financial resources and a permanent population with a diversity of human skills. Arctic laws, regulations, resource management policies and plans are effective in proportion to the availability, capacity and adequacy of local infrastructure, which is adequate in many regions, but not all. Not surprisingly, the poorest Arctic areas can muster more management infrastructure than the Antarctic – which has no financial resources, no permanent population and virtually no infrastructure.

A considerable hazard facing polar mariners is incomplete or inaccurate charts, rendered questionable by climatic changes that alter coasts and land forms. Canadian officials warn that, while melting sea ice has drawn more cruise ships to Canada's Arctic waters, maritime charts that date back to the Arctic explorer, Sir John Franklin, in the early 1800s may no longer be accurate (CBC News, 4/16/08). New reefs, shoals, even islands appear as glaciers retreat: tourists may enjoy viewing the drama of calving glaciers, while navigators are unsure of the depth of water ahead.

Arctic nations are building new facilities and streamlining procedures for dealing with incidents. Sweden and Russia, for example, are building and commissioning new coast-guard vessels. Finland has built a fleet of 15 oil-spill response vessels, though lack of crews has made it possible for only 10 to operate. The Danish Navy is providing very competent emergency response skills for Greenland, but does not have sufficient vessel capacity to evacuate the large numbers of passengers and crew aboard the cruise ships now visiting the Greenland coast. Iceland has consolidated its emergency response services nationally and offers a fine model for well-integrated emergency response.

Canada's Coast Guard manages its waters very effectively, but expresses concerns about future volumes of ship traffic. In 2007, Canada created a new Pan-Territorial Northern Search and Rescue Strategy that brings together the Emergency Measures Organizations and the Royal Canadian Mounted Police Divisions of the Northwest Territories, Yukon and Nunavut, to focus on search and rescue exercises, training and prevention issues specific to the north (Government of Northwest Territories, 2007).

The US Coast Guard, by contrast, does not have sufficient vessels to conduct regular Arctic patrols. In 2008, the commandant of the US Coast Guard pointed out that less ice means more ships making their way through formerly impassable waters, and will require more search and rescue missions and oil-spill responses. More icebreakers will be required to keep up with this demand (*Wired*, 4/14/08).

Mutual aid and joint training exercises help to improve response capabilities. Icelandic and US Coast Guards have agreed to cooperate on rescue operations (*Iceland Review*, 14/18/07). The West Nordic Council has recommended that the governments in Iceland, the Faroe Islands and Greenland strengthen their collaboration with other Nordic and North Atlantic countries, particularly on emergency response and coordination (Arctic Council, 8/28/07). Since 2001, Sweden, Norway, Finland and Russia have conducted joint Barents Rescue exercises to prepare for emergencies that may occur in the Arctic Ocean (*Barents Observer*, 9/4/07). Norway has decided to replace 10–12 of its helicopters by the year 2015, and has teamed with Iceland to consider a joint procurement process. Both countries are members of the NAWSARH project (Norwegian All Weather Search and Rescue Helicopter), which is organising the acquisition. The project expected to release a tender in 2009 (*IceNews*, 2/22/08).

The successful evacuation of passengers and crew from stranded or sinking vessels is completed when they arrive at shore-based facilities and services that can shelter, feed and provide medical treatment. Obviously, the capacity of small, remote Arctic communities to perform these functions for several thousand cruise ship passengers is severely burdened by a shortage of facilities, trained personnel and provisions. As a consequence of the growing size and numbers of Arctic cruise ships, it is glaringly evident to Arctic communities that their infrastructure capacity will be overwhelmed in the event of an emergency.

Despite current efforts to improve the Arctic's infrastructure capacity and its ability to respond to marine incidents, there is much work to be done in exceedingly difficult environmental conditions. Information, facilities, equipment and human resources barely meet the Arctic's current volume of shipborne traffic, and infrastructure will have to expand considerably to keep up with future needs. The complete lack of any such facilities in the Antarctic, despite a rapidly growing number of visiting ships, is now a matter for serious concern.

For recent stateholder comments on deficiencies in regulation of Antarctic tourism see Haase and others (2007, 2009).

Adventure Mariners

Thousands of visitors tour polar regions, especially in the more accessible Arctic, aboard private motor yachts, sea kayaks and sailboats (Figure 6.2). Such adventure mariners may or may not be acquainted with the unique environmental features and hostile conditions they are likely to encounter. When their skills and knowledge are insufficient to cope fully with polar conditions, they place both themselves and the environment at risk, and become unwanted candidates for overstretched search and rescue operations.

Independent marine travel in regions with unreliable telecommunication services may create dangerous situations to which Arctic governments can respond only with difficulty. Search and rescue agencies and those responsible for environmental protection rarely know the location or schedules of adventure mariners, and have no knowledge of their skills or the adequacy of their provisions. This is especially true of sea kayakers, who may travel widely, camping, cooking and sleeping ashore at different locations, and for whom a seemingly small accident can turn into a major rescue effort. As more such adventurers perceive that both polar regions are becoming more

Figure 6.2 Are recreational boaters prepared for the polar seas?
Photo: JMS.

accessible due to warming, this small but hazardous market segment will probably continue to grow.

Summary and Conclusions

Marine tourism, the fastest-growing segment of polar tourism, seems likely to continue growing, helped by pressures both from clients and from government agencies and service industries that profit by supplying it. Operating in hazardous conditions, it remains an industry subject to dangers that require special management policies and resources, many of which it has already outgrown in the north. We cite examples of near-disasters involving polar cruise ships, and consider the threats to human and environmental safety that they imply.

We review some of the legislation under which ships currently operate, concerned mainly with protecting human life and the environment. Human safety, quite properly the first consideration for sound management, is at particular risk due to distances of polar ships from normal shipping lanes and sources of help. Rapid response to emergencies is essential, but not always practicable: the most likely source of help is usually another cruise ship in the same area. Environmental safety is challenged by the very presence of ships in highly sensitive areas. We draw particular attention to the effects of oil spills, and problems of sewage, wastewater, bilge and solid waste disposal where facilities for dealing with them are limited. Air pollution from smoke-stack emissions are also considered, and the hazards to wildlife presented both by ships at sea and passengers on shore excursions, again in areas of particular sensitivity.

We review some of the guidelines provided for their own use by associations of companies in both the Arctic and the Antarctic, and consider requirements of infrastructure to be provided by national authorities. We finally draw attention to the special case of adventure mariners, who operate independently, often out of sight or knowledge of authorities, but are as likely as other tourists to meet difficulties and require help or other services from time to time.

Chapter 7
Polar Culture and Heritage Tourism

Introduction

> Mush your own dog sled through the Arctic wilds of Norway and Sweden, staying with a Saami family along the way.

So reads an advertisement for a 2009 Abercrombie and Kent Arctic tour, combining adventure, culture and heritage. Culture and heritage tourism is a market sector for tourists who want to visit historic places, experience personal interaction with the lives and traditions of native people, or simply learn more about historical topics that interest them. More often associated with temperate or tropical sites of ancient civilisations, it is nevertheless a reason given by many travellers for heading to polar regions.

Arctic culture and heritage are based on the native peoples whose nomadic ancestors first colonised the northlands some 30,000 years ago, and their successors who spread northward as the land ice receded (p. 8). Still later came the colonisers who farmed Iceland and southern Greenland, the missionaries, and the 18th- and 19th-century merchant adventurers who exploited the Arctic's whales, seals, fur-bearing mammals and minerals (Chapter 2). Most recently, the arrival of scientists, administrators, public servants, contractors and military has brought the 20th century to the regions. Each group established cultures and traditions, leaving heritage signatures that range from stone-circle campsites to warehouses, forts, monasteries, abandoned prison settlements and DEW-line installations. Most indicative of the future are the towns, townships and settlements of the modern Arctic.

Antarctica and its neighbouring islands appear never to have supported native populations: the region shows a heritage of sealing, whaling and exploration relics, but little culture dating before the 20th century. Whether in the north or south, cultural traditions and heritage sites require sensitive handling in the presence of tourism. This chapter discusses what history and culture are on offer for modern-day tourists, and how they are managed at either end of Earth.

Arctic People and Communities

Today, approximately 4 million people live in the Arctic. Of these, about half live in the Russian Federation, four fifths of them in sizeable cities. Though none of the other seven Arctic nations has industrialised its northlands or concentrated its Arctic populations to a comparable

extent, each has at least one town-sized administrative centre (Table 7.1), several smaller townships and a range of widely scattered rural settlements that house the previously nomadic populations.

Table 7.1 Populations (c. 2005) of Arctic cities and towns of the eight Arctic nations

City	***Population***
Canada	
Whitehorse	20,500
Yellowknife	18,700
Finland	
Rovaniemi	60,000
Greenland	
Nuuk	15,000
Iceland	
Reykjavik	195,000
Akureyri	17,300
Norway	
Tromsö	53,500
Russia	
Murmansk	336,000
Norilsk	135,000
Vorkuta	85,000
Noyabrsk	100,000
Novy Urengoi	96,500
Sweden	
Luleå	45,500
USA (Alaska)	
Anchorage	280,000
Juneau	30,500
Fairbanks	34,500

Source: Wikipedia and other sources

Only about 10% of the Arctic population overall can claim to be 'indigenous' in the sense of 'locally born of original stock'. For various interpretations of 'indigenous' used in different administrations, see Box 7.1. Few that qualify in any sense of the word retain the traditional nomadic ways of life at subsistence level (see Chapter 3). All Arctic inhabitants are today full citizens of the countries in which they find themselves, with full civil rights and responsibilities, and all share at least some of the comforts and advantages of modern life. Most live in houses, send their children to schools, use motorised transport, receive pay packets, buy their food, clothing and furnishings in supermarkets and are nominally within reach of efficient medical care.

Box 7.1 Arctic populations

Not all Arctic countries recognise either 'Arctic' or 'indigenous' in their official statistics. According to Bogoyavlenskiy and Siggner (2004: 28), Greenland statistics distinguish individuals born in Greenland from those born outside: about 88% of the population is, in this sense, indigenous Greenlanders. Saami ethnicity is not registered in Norway, Sweden or Finland, but about half the population of these countries' Arctic regions are considered to be Saami (Arctic Monitoring and Assessment Programme, 1998: 161). The Canadian census distinguishes Inuit, North American Indians and Métis as indigenous. In the province of Nunavut, the proportion is 85%, in the Canadian Arctic as a whole, the proportion is about 50%. In the USA, indigenous people include American Indians and Alaskan natives, making up 15–20%. Under the Alaska Native Claims Settlement Act, individuals may enrol as natives if they are one quarter indigenous (Young & Einarsson, 2004: 239). The Russian census recognises 13 indigenous peoples, who together make up roughly 4% of Russia's Arctic population.

About two thirds of the pan-Arctic population live in settlements of over 5000. This proportion also varies widely between countries, ranging from over 80% in the heavily industrialised Russian north to over 60% in Alaska and only about one third in Greenland. Populations throughout the Arctic grew rapidly from the 1950s, but later stabilised and have recently tended to remain static or decline. Natural increase is generally higher among indigenous people than non-indigenous in the same areas. According to Gorman (2005: A-6), native populations of northern Canada have the fastest rate of growth in Canada and among the fastest in the world, increasing at a rate of 16% per decade.

Transition from traditional to modern life has not been easy for all. The rapid pace of changing cultural practices and community values has resulted in mental health problems in indigenous people, including depression, violence and high rates of suicide (Hild & Stordahl, 2004). Young and Einarsson paint a brighter picture. While accounts of such problems are not to be dismissed lightly:

> they do not tell the whole story regarding human development in the Arctic. Many individual Arctic residents are highly successful. There are sharp differences among the communities of the region with respect to any number of measures of viability. The Arctic's residents have a long history of successful adaptations, even in the face of rapid and far-reaching changes of the sort that are impacting them today. (Young & Einarsson, 2004: 236)

The same authors record the conspicuous success of Arctic communities in maintaining cultural integrity, adopting such modern technologies as telemedicine and distant education, and the development of innovative systems of government, all in the face of a variety of external pressures. Among those pressures, for better or worse, may be included the advent of tourism – in particular, mass tourism that, if uncontrolled or poorly managed, brings large numbers of strangers unbidden to the doorsteps of hitherto remote settlements.

Seasonal visitors to the Arctic now outnumber host populations at all popular destinations, and thousands of tourists arriving *en masse* may easily exceed infrastructure capacities and inundate social institutions. Ill-managed culture and heritage tourism at its worst can be highly intrusive and disruptive. Well-managed and at its best, it may provide a lifeline of enterprise, bringing new interest and opportunities for enterprise to settlements that are short of both. It may provide social bonding that helps to hold communities together when fishing, hunting and other traditional use of resources are failing.

Arctic Culture and Heritage Resources

Resources of interest to culture and heritage tourism (Figure 7.1) can be listed under three headings:

- *Culture*: concerned mainly with people in contemporary communities. Many Arctic communities, both large and small, contain a proportion of indigenous folk who are concerned to preserve at least tokens of their native languages, ceremonial customs and practices and artistic expression, and to share them with visitors.
- *Heritage*: sites and structures concerned mainly with history of communities and cultures, cherished as evidence both of former cultures and of successful adaptation to change. These too may be

Figure 7.1 Arctic heritage is an attractive tourist venue. Colonial Russian churches in Arctic Alaska. Photo: JMS.

shared with visitors, usually in supervised parties under local guides who provide written hand-outs or spoken commentaries.

- *History*: artifacts derived from the history of Arctic exploration and development. Historic relics that lie close to existing communities may be taken over, protected and maintained for display as tourist attractions. Isolated examples that are remote from settlements may lack any form of guardianship. Previously protected by their isolation, as tourism grows they are increasingly open to abuse and vandalism.

The first and second are concerned mainly with Arctic people, their history, how they lived in the past and how they live today. The third – purely historical artifacts – are derived from the efforts of outsiders to penetrate the Arctic for exploration or exploitation. They include, for example, the graves of naval seamen, whalers, sealers and trappers scattered along the shores of Svalbard, Greenland and the Northwest Passage; try-pots and other shoreside whaling relics; abandoned mine workings and military stations, and the huts, hangars, masts and long-abandoned equipment of expeditions that explored by land, sea or air.

Changes in administration

Throughout the Arctic these resources have until recently been held in the care of distant southern-based administrations. Governments in Washington, Ottawa, Oslo and Moscow have designated protected areas, allowed or prohibited hunting of particular species, declared national parks and invited in visitors, all with minimal reference to the native inhabitants. Such remote control has not always worked well. Attempts by the tourism industry to work amicably with indigenous people have been frustrated; attempts by governments and tour operators to develop tourist projects have proved unacceptable to local people, who felt that their needs were not sufficiently taken into account.

This situation is changing. As northern populations have grown, southern governments have found it expedient to create northern jurisdictions that provide credible rôles for indigenous people in the management of their own cultural and natural resources. Alaska Native Corporations, Greenland Home Rule Government, Nunavut Government in Canada and the Saami Circumpolar Council have all proved successful in directing the management and allowable uses of their resources (see below).

From the new administrations are arising well-designed culture and heritage programmes that are considerable improvements on what has gone before. Programmes that, for example, help to preserve traditions, cultural values and use of languages (Amberger, 2003; Milne *et al.*, 1995) now operate in Alaska, Scandinavia, Nunavut, Greenland, the Faroe Islands and Iceland. The trend is spreading: currently some 42 indigenous groups, comprising the Russian Association of Indigenous Peoples of the North, are determining how their cultures can be made attractive to tourists (RAIPON, 2007).

Changing economic trends

Among the economic changes that currently affect the northlands (Chapter 3), three in particular most directly affect tourism:

- increasing marine shipping;
- limitations to commercial and subsistence hunting, herding and fishing;
- expansions of mining and oil and gas extraction.

Increasing marine shipping implies more visits from cruise ships – advantageous to ports with the infrastructure to deal with their passengers, less so to those without, and not always environmentally benevolent (see Chapter 6). Reductions in commercial fish catches cripple the industry that is currently the main source of income for hundreds of Arctic towns and settlements. Limitations to subsistence hunting reduce incomes and standards of living in the smaller settlements. Either trend may be countered by increasing and diversifying tourism. The mining and hydrocarbon industries have provided jobs where they are needed. However, mines have opened and closed at the bidding of world markets, without reference to local needs. Hydrocarbon extraction has proved more stable, and is likely to continue into the future. It requires mainly skilled labour, and may progressively employ more local people as more acquire the necessary skills.

Reserves of oil and gas underlying Arctic tundra and coastal shelves are viewed hungrily by consumers worldwide, and likely to be brought into production as existing fields become exhausted. Arctic wildlife already finds itself sidelined when oil and gas are needed, and tourism that depends on wildlife in wilderness settings will fare no better. However, tourism is an opportunist industry. In Alaska, northern Canada and Siberia, the infrastructure of ports, roads and settlements required by other industries has in the past provided opportunities for tour operators to acquire new venues. If devastated tundra, Gulag prison sites, abandoned mine workings and derelict military installations already prove viable as tourist attractions, there must surely be a future for tours of oil and gas installations.

From its rate of acceptance by local communities, culture and heritage tourism appears to provide more sustainable incomes than alternatives that are at present available. Communities that operate within self-rule especially seem able to provide jobs and create markets for their products and services. They approach the industry in a variety of ways. Snyder writes:

> Some have formed corporations and councils, such as the Alaska Native Council, that actively seek to finance and promote this form of development. Others, such as the Nunavut, have exercised governmental powers to implement this form of tourism. Still others, such as the Saami, employ smaller-scale, community-based lodgings to share their traditions and culture... Governments throughout the Arctic assist these rural economic and cultural development efforts by means of legislation, financial subsidies and promotional campaigns in support of native arts and crafts. (Snyder, 2007: 65–66)

Small-scale Saami accommodation does not mean catering for small numbers of visitors. There are only 4000 Saami in Finnish Lapland, but individual lodges have each year hosted 130,000 day visitors and 1200 overnighters (Rennicke, 2004).

Managing Arctic Culture and Heritage Tourism

Tourism that brings clients to where Arctic people live, but fails to take account of their values, is both inept and self-destructive. Successful management is aware of these values, and takes steps to ensure that they are (a) willingly offered by the Arctic folk themselves; and (b) protected by law, regulation, convention and every other possible means, to secure continuity of respect and continuity of use (Box 7.2).

Box 7.2 Misunderstandings

Social anthropologist Mark Nuttall provides interesting examples of how European environmental concepts of 'development' in northeast Greenland fail to mesh with those of Inuit communities in northeast Greenland. A proposal to open a seal-meat canning factory in one community (Nuttall, 1992: 145–147), for example, was met with scepticism by the local hunters' association:

> and criticised for lack of understanding on the part of the municipality. The hunters felt that both Greenlanders and Danes involved with bureaucracy in Upernavik had failed to grasp both the realities and meaning of seal hunting... the idea of selling meat goes against the ideology of subsistence. Because seal hunting continues throughout the year and... is done to provide the hunter's family with meat, it does not expect any direct material return.

Thus, community values are placed at risk if a commodity that is readily shared among kinfolk and neighbours is bought and sold. In the village, seal meat freely given is thus inalienable, containing an element of the giver, and with value extending far beyond nutrition and maintaining life:

> As a gift, seal meat symbolizes cultural continuity and sustains close social networks... if meat becomes a commodity it can actually harm social relationships and the cultural fabric of community.

By contrast, the autumn beluga (white whale) hunt is seasonal, like the halibut fishery. Both beluga meat and fish have recognised commercial value, and can equally be bought and sold to strangers.

It would clearly not be difficult for bureaucrats, deficient in knowledge of local sensitivities, to make similarly well-intentioned but socially unacceptable suggestions concerning tourism in local communities.

Direct negotiations between local folk, local governments and tour operators has helped to strengthen Arctic cultures, allowing native people to share them with less risk of violating integrity or privacy. Cultural groups that want to participate in polar tourism, whether Inuit, Saami or Alaskan Native, determine what aspects of their cultural resources will be shared, how they can authentically be represented, and where and when tourism may occur. Strengthening is manifest in the resurgence of traditional languages and customs, initiatives taken in passing favourable legislation, protection of heritage sites and the wide acceptance of culture and heritage tourism by native communities, to their own benefit and to the pleasure and instruction of thousands of visitors.

Though the primary image of 'culture and heritage tourism' may be visits to small communities that demonstrate aboriginal ways of life, culture and heritage are presented in many other ways:

- Cities and towns provide opportunities for visitors to discover a wider range of both through museums, theme parks and craft centres, covering not only current ways of life, but also the succession of cultures leading up to the present.
- Cruise ships visit remote islands of the Canadian Arctic archipelago – e.g. Beechey Island, where graves, cairns and other artefacts dating from Sir John Franklin's ill-fated expedition (1845–1846) and others of the period can be seen.
- Regular scheduled cruises to Svalbard include opportunities for dog-sledging, a visit to a coal-mining museum, exploring remnants of 18th-century Dutch and British whaling stations and visiting remote 19th-century trappers' huts.
- Steam trains and ships carry passengers in comfort to Kola Peninsula and the White Sea, to visit monasteries and – if they are so minded – industrially devastated tundra and a Gulag prison site.
- Trains, ships and coaches provide sophisticated travel to remnants of Alaska's 19th-century gold rush in Fairbanks and Dawson City.

Visitors may be limited to buying goods and services, or invited to watch or participate in selected activities: examples of both appear in reports by Walle (1993) and the Alaska Native Council (2005). Visiting is one thing, appreciation another; culture and heritage tourism more than any other branch of the industry requires sound briefing and guiding, to ensure that clients gain full benefit from their experience. 'It's Friday, so we must be in Anchorage' may characterise a vacuous tourist; it can also indicate the missed opportunities of inept management.

Determining ways in which this can best be achieved may involve difficult choices. Allowing visitors to visit communities and participate in traditional and ceremonial practices has to be balanced against protecting privacy and community values, and take into account alternative uses of local resources. Huntington *et al.* (2007) provide examples of choices made and their consequences, in case studies of interactive cultural tourism at three localities in rural Alaska:

- At Anaktuvuk Pass, a last-remaining enclave of inland Iñupiat in the North Central Brooks Range, visitors are guided around the village and an excellent folk museum, introduced to local people and treated to a display of dancing. Many opportunities exist in this attractive locality for further tours and displays.

- At Kotzebue and other localities in northwest Alaska, a major tourist attraction, successfully pursued, is wildlife hunting with local guides. However, it has created problems for local subsistence hunters, and 'tourism remains an activity imposed from afar rather than one embraced and engaged locally' (Huntington *et al.*, 2007: 78).
- In Yakutat Bay, a huge increase in numbers of cruise ships (from 15 in 1993 to over 160 by 2007) has disturbed breeding seals on which local hunters relied for meat and oil. The city of Yakatut received funds to cover some of the costs of providing tourism and recreation infrastructure and emergency services, and the Yakatut Tlingit Tribe received money from the industry to fund professional research on seal disturbance. The Northwest Cruise Ship Association is providing five additional years' payments to the tribe, a portion of which will fund further research.

Robbins (2007) illustrates similar problems in helping Inuit communities of the Canadian Arctic to develop eco-lodges (accommodation for travellers interested in combining studies of indigenous culture with adventure, wildlife viewing and archaeology) and other tourism ventures. At Pangnirtung, Baffin Island, Robbins assisted in developing a wide range of government-funded tourism facilities, including an interpretive centre with museum and elders' room, a lodge, a refurbished former whaling station and a range of tourism facilities from craft shops to hiking trails. By 1999, these were providing 35 permanent and 14 seasonal jobs, and income for over 100 home-based artists (Budke, 1999).

From a wide range of experience in such developments, Robbins comments:

> The Inuit in Nunavut... are interested in tourism as a form of economic development and employment, but they are also still concerned with community control to minimize the intrusive nature of tourism. There is a need... for the Inuit to be further involved in the tourism supply chain, as most tourists coming into the territory are being packaged by southern companies. (Robbins, 2007: 100)

Robbins concludes that keys to future growth of tourism in Nunavut lie in:

- more and enhanced training opportunities for workers in the sector;
- better community and political awareness of the benefits of tourism;
- access to capital;
- development of more export-ready products and experiences;
- ultimately more control of tourism in Inuit hands.

These conclusions are likely to be shared by all whose interests lie in promoting culture and heritage tourism for the benefit of local communities throughout the Arctic.

Antarctic Heritage Tourism

Culture and heritage are not the first topics that come to mind as motives for visiting the Antarctic region. Nevertheless, tourists to Antarctica generally admit interest in the region's relatively short history, particularly its history of early 20th-century exploration, and are duly impressed on visiting some of the few artifacts that remain from that period.

The earliest of all human artifacts in the region are 19th-century sealers' camps on islands of the Scotia Arc. Next in age are remnants of early 20th-century whaling installations on some of the peripheral islands. Continental Antarctica's oldest buildings are the huts of the British Antarctic (*Southern Cross*) Expedition 1898–1900 at Cape Adare, Victoria Land. A few later expedition huts and sites from the early 20th century remain as evidence of exploration and scientific inquiry – the main human incursions at that time.

Since 1972, the Antarctic Treaty System has recognised the need to preserve the region's few historic relics. In 1991, Appendix V, Article 8 of the Protocol on Environmental Protection to the Antarctic Treaty formally provided for the listing of Historic Sites and Monuments on the continent and islands within its jurisdiction (i.e. south of 60°S latitude). The list currently (2008) includes 82 sites (Box 7.3), and is augmented from time to time by Treaty parties. Under the Protocol, listed historic sites and monuments 'shall not be damaged, moved or destroyed'. However, as Hughes and Davis point out:

> There are no established criteria for determining historic sites and monuments, no clear philosophy about "conservation", "restoration", removal or "interpretation" and no Antarctic Treaty-recognised guidelines. (Hughes & Davis, 1995: 235)

Since this comment was written, recognised guidelines for behaviour of tourists in the Antarctic generally have been introduced and accepted (Chapter 8), but the Treaty parties have done little else toward establishing either the criteria or the clear philosophy that Hughes and Davis seek. Hughes & Davis' (1995: 240) warning that, 'it is a matter of urgency that management strategies for historic huts and other cultural sites be developed', is perhaps wishing too much responsibility on an organisation that has more pressing problems to solve. Responsibilities for upkeep and maintenance rest with the state or states that propose the site. Hughes and Davis continue with practical suggestions on documentation, conservation and interpretation, and measures for dealing with tourism at historic sites and monuments (see below Box 7.3).

Box 7.3 Historic sites and monuments

The Antarctic Treaty register of historic sites and monuments puts on record places and human artifacts that governments regard as being worthy of preservation and protection from damage. To be recognised on site, monuments are required to be marked with notices in English, French, Spanish and Russian, the four official languages of the Treaty. A full list of sites is available on the Antarctic Treaty website (http://www.ats.aq/index_e.htm).

Of 82 sites registered by 2008, over one quarter are huts and shelters dating mainly from early 20th-century expeditions. They include many that are easily visited by tourists on scheduled cruises (see below). Others, for example the remains of the stone shelter rigged by Edward Wilson's party on their Cape Crozier journey (Huxley, 1914: Vol. 2), lie far from normal tourist routes. A few recent huts dating from the 1940s and 1950s are included. The state or states that nominate them must undertake to maintain them. Huts that are disused and not listed must be removed – a ruling that in recent years has mercifully rid Antarctica of many derelict post-WWII expedition huts and accompanying rubbish.

The list includes a single whaling station (at Whalers Bay, Deception Island), of which little more than iron boilers, oil tanks, winches and fragments of buildings remain. Sadly, the customs officer's residence, which was intact up to the 1980s, lost its roof to a gale and is now barely recognisable – a reminder that listing a building in a schedule is not in itself protection against Antarctic wear-and-tear.

One curious entry among the shelters is 'No. 80, Amundsen's tent', which was erected at the South Pole on 14 December 1911. Visited by Scott's party on 17 January 1912, it has since disappeared, and presumably lies deep under the snow and rime of the polar plateau. Norway has registered it as a historic site and monument. Should anyone find it, the 1991 Protocol affirms that the tent must not be damaged, moved or destroyed.

More tangibly, the list includes graves of expedition members, memorials to those whose remains were not recovered, and a selection of commemorative cairns, poles, masts, plaques, crosses, monuments and busts. A cairn on Half Moon Beach, Livingston Island, South Shetland Islands, commemorates the earliest event recorded. Nominated jointly by Chile, Spain and Peru, it honours the passengers and crew aboard the Spanish vessel, *San Telmo*, which is belived to have sunk nearby in 1819.

What is and is not 'historic' is clearly an issue too controversial for an international forum to tackle with any hope of agreement. Warren (1989: 89, quoted in Hughes & Davis, 1995: 240) considered such a site to be:

> Any place which was the location of, or is fundamentally associated with, an original and significant event in Antarctic discovery, exploration or science prior to the IGY [and] represents a unique international achievement [or] symbolises an important artistic, national, religious or cultural event.

Regarding the International Geophysical Year (IGY) 1957–1958 as the end-point of the historical period seems sensible enough: it was indeed the point at which human attitudes to Antarctica changed radically, as manifest in the Antarctic Treaty that emerged three years later. Most of the sites so far designated fit well into this time frame. Whether they all represent significant historical events in the history of the continent is another matter: there are still no more rigorous criteria for inclusion than the desire of a Consultative Party to make a nomination.

More historic sites are found on some of the sub-polar islands that lie beyond the limits of the Arctic region as defined in this study (Chapter 1), and are often visited by cruise ships on their way to and from Antarctica. Unlike Antarctica itself, these islands are controlled and managed under the sovereignty of individual states that accept full responsibility for them, including protection of their few cultural, heritage and historic sites. The different management practices that arise from this distinction are discussed in Chapter 8.

Sealing and whaling sites

The Antarctic region's earliest human visitors were late 18th- and early 19th-century sealers who, following indications given by Captain James Cook RN after his voyages of the 1770s, stripped South Georgia of its fur seals, then extended their search to other islands of the Scotia Arc. Through the first half of the 19th century, sealing parties stayed ashore in accessible coves, killing fur seals for their pelts and elephant seals for their blubber, which was an acceptable alternative to whale oil. They brought little ashore, apart from sailcloth and spars for shelter, and cast iron try-pots, set up on brick hearths, in which they rendered the seal blubber. Spars, fragments of canvas, bricks, try-pots, clay pipes, bottles and a scattering of graves, are the sparse remains of their industry. Several such sites have been identified on South Georgia (Headland, 1984) and the South Shetland Islands: more are likely to exist under invading coastal vegetation. Though landings from cruise ships may be made nearby, they do not feature on tour itineraries, and neither the

Figure 7.2 Historic relics at whaling stations are simultaneously attractive and dangerous. Photo: JMS.

Antarctic Treaty nor claimant governments provide information about them or their origins.

More evocative as heritage sites are the whaling stations (Figure 7.2), established from 1904 onward in sheltered harbours along the north coast of South Georgia, on Deception Island in the South Shetlands, and on Signy Island, South Orkney Islands. For comprehensive histories of the whaling industry on South Georgia and in the Antarctic Peninsula area, see Hart (2001, 2006).

Of six stations built on South Georgia, one closed in 1920 and a second in 1931. The four remaining stations – Stromness, Leith Harbour, Husvik, Grytviken – closed during the mid-1960s. The buildings were left intact and 'mothballed' by their owners, but were quickly looted and vandalised, mostly by visiting fishing ships. Though sturdily built of Norwegian timber, all have quickly deteriorated in the damp, windy climate. It was still possible for shipborne tourists to walk through them during the 1980s and 1990s, but continuing deterioration has eventually made them all unsafe, particularly in high winds when sheets of corrugated iron blow from the roofs and walls.

At Grytviken, oldest of the stations, a study by the Norwegian Antarctic Research Expedition produced comprehensive inventories, industrial archaeological surveys and documentation of the station. Working parties subsequently stripped out all dangerous materials, and shipborne tourists can now walk freely through a sanitized version of the station (Basberg, 2004). The former Manager's House has been converted to an excellent museum, which tells the history of whaling from South Georgia and throughout the South American sector. Close by is the station graveyard, which holds one further and major tourist attraction – the grave of British explorer, Sir Ernest Shackleton.

Today, little is left of the Deception Island station, which closed in 1932. It remained almost intact until the mid-1940s, but was severely damaged by a mud-slide during volcanic eruptions in the 1970s. The site of the small Signy Island station, which operated for only a few years in the early 1920s, has been almost completely usurped by a British scientific station. Many other relics of the industry can be found in the Peninsula area, including the wreck of a transport ship burnt out in 1916, massive factory-ship moorings in sheltered bays and harbours, and remnant stockpiles of coal and oak barrel staves, now often adopted by nesting petrels.

Historic huts and shelters

For many visitors, the most evocative of Antarctica's historic building are the living huts of expeditions of the so-called 'heroic age', an ill-defined period of exploration covering the first two decades of the 20th century. Next oldest after those at Cape Adare (p. 131) is the 'Discovery' hut of Robert Falcon Scott's British National Antarctic Expedition 1901–1904, in McMurdo Sound, Ross Dependency. Named for the expedition ship that wintered for two seasons close by, the hut when built was the world's southernmost and most isolated human dwelling. Though in accord with the Protocol, it has suffered neither damage, removal nor destruction, it is now almost eclipsed by the neighbouring US McMurdo research station. On capes further north along the Ross Island coast are the huts of Shackleton's British Antarctic (*Nimrod*) Expedition 1907–1909, and Scott's British Antarctic (*Terra Nova*) Expedition 1910–1913, both mercifully still in isolation.

Older than these is the wooden hut of Otto Nordenskiold's Swedish South Polar Expedition 1901–1904, on Snow Hill Island, Antarctic Peninsula. In the same area are the remains of two contemporary stone huts, built by members of the same expedition after their ship, *Antarctic*, was crushed by pack ice and lost. A small hut at Hope Bay, just big enough to contain the three men who wintered in it, now stands at the crossroads of the substantial modern Argentine station, Esperanza. A larger hut on Paulet Island, built by the 22 men who survived the shipwreck, and originally roofed with spars and sails, remains in derelict condition surrounded by penguin nests, but as yet unencumbered by later buildings.

At Cape Denison, Commonwealth Bay, on the coast of George V Land, stands the lone hut of Douglas Mawson's Australasian Antarctic Expedition 1911–1914. Though built with no intention of being used for more than the two years of the expedition, it has withstood almost a century of near-continuous blizzards. Its timber fabric is now worn thin by erosion, its future a matter of concern.

All the surviving timber-built huts on mainland Antarctica stand up well in the cold, dry climates. Those in the damper Peninsula area deteriorate faster. Huts erected during the politically motivated building boom of the 1950s and 1960s and abandoned when no longer needed have quickly become derelict and a liability to their builders.

The truly historic huts, most of which appear on the List of Historic Sites and Monuments, are visited several times yearly by parties of shipborne tourists. Visitors seldom fail to be moved by their homeliness – the bunks, blankets, worn and blubber-stained windproof clothing, canned food, dried-out hams, books, magazines, dog-harnesses and bales of horse-fodder (the latter still emitting a faint farmyard smell when the sun strikes them on warm days). However, they were not built to last, and their gradual but inevitable deterioration poses management problems. They will not survive indefinitely where they stand. Should they be encased in protective domes? Or brought back for re-erection in museums?

The first problem, to whom do they belong and who accepts responsibility for them, was solved by the claimant governments (New Zealand and Australia) taking them over and assigning responsibility to licensed charitable heritage trusts. Limited tourist access is granted, but only by previous arrangement and in the presence of authorised key-holding guides. These actions might be questioned under the terms of the Antarctic Treaty (perhaps under Article IV(2), as actions constituting an enlargement of an exisiting claim). However, those aware of the reality of the situation regarded it as a triumph of common sense: the huts were under threat of deterioration – someone needed to take charge, and someone did. The remaining problems of subsidising their upkeep and deciding their ultimate fate are, as yet, unsolved.

A more recent historic hut, at Port Lockroy, Wienke Island, off the Danco Coast of Antarctic Peninsula, dates from a later period of exploration. Built in 1943–1944, it was the first expedition hut of 'Operation Tabarin', a British expedition of WWII that became the forerunner of the Falkland Islands Dependencies Survey and ultimately of British Antarctic Survey (Fuchs, 1982). Though long abandoned, it was recently refurbished to become a museum for visiting tourists. Including a small shop, it is currently reputed to be Antarctica's most popular tourist landing site.

American tourists visiting Antarctica, noting with interest the British expedition huts, are concerned to know why those of Roald Amundsen and Richard Byrd are not in evidence. The answer is that both Amundsen and Byrd built their stations (in Byrd's case, five successive versions, which he called 'Little America' I to V) on the Ross Ice

Shelf, where they were gradually encased in ice and – as the Shelf advanced – have ultimately been deposited in the Ross Sea. The single exception is a station built under Byrd's command on Stonington Island, Marguerite Bay, Graham Land, for the US Antarctic Service Expedition 1939–1941. Designated HSM 64 and refurbished in 1992 (Broadbent, 1992; Spude & Spude, 1993), East Base has since deteriorated to the point of dereliction.

Antarctic place names

In a region where history is short and culture thin on the ground, maps and charts spring to prominence for their record of history in place names. Every navigator from the 18th century onward gave names to the features that he entered on his charts. Later navigators, unaware of earlier ones, gave new names to features that already had several assigned to them. The same name was often applied in several languages to many features: there were, for example, innumerable 'Penguin Islands' and 'Seal Points'. Matters came to a head in 1948 when Byrd's expeditions following WWII demanded a plethora of new names. This demanded a complete revision of the names themselves, and a coordination of the systems by which names are bestowed.

Each interested state established or reconvened an Antarctic Place Names Committee, and the various committees compared notes and reached agreements. There resulted several national reviews of place names, and ultimately an international comprehensive two-volume, *Composite Gazetteer of Antarctica*, produced by the Working Group on Geodesy and Geographical Information of the Scientific Committee on Antarctic Research (SCAR, 1998). Between them, these publications sorted all place names up-to-date, providing a firm foundation to which new names can be added as new places are identified.

Antarctic place names include many clues to their origins, most of which were the expeditions on which the particular features were identified and named. This is particularly well illustrated in the Antarctic Peninsula region, where German, Belgian, French, Swedish, Scottish, Norwegian, Argentine, Chilean and American names reflect the expeditions from which they originated. The most comprehensive guide to their origins is Hattersley-Smith's (1991) two-volume, *History of Place-names in the British Antarctic Territory*. For Antarctica as a whole, see the more bulky but more comprehensive, *Geographic Names of the Antarctic* (Alberts, 1995). Both provide excellent keys to understanding this particular aspect of Antarctic heritage – how, when and by whom the Antarctic region south of 60°S was discovered, mapped and named. Similar guides exist for South Georgia and other fringe islands.

Managing Historic Sites for Tourism

During visits to Antarctic wildlife sites, tourist parties usually spread over a wide area. At historic sites they remain concentrated; their potential for damage is greater, and more rigorous guidelines or rules are needed to ensure minimal impacts on the sites. Very little attention has been paid to the practicalities of site management. Hughes and Davis (1995: 250–253) have elaborated 'elements of an action plan' for managing tourism at such sites, covering a wide range of topics and possibilities. In our view, the following three points express what is needed first in Antarctic field situations, on which more detailed action plans for individual sites can be developed.

Management of an historic site requires:

- Acceptance of responsibility for site management. As the Antarctic Treaty System accepts no practical responsibility, it must be delegated to the listed 'parties undertaking management', usually the state or states that have nominated the site.
- Preliminary documentation before visits are allowed. What and where is the site? How is it marked? What are its significant points? Which parts of the site are to be available for visits and which restricted? At present, none of this seems to be required: all of it is essential in defining a site if protection is to be taken seriously.
- A management plan, emphasising the practicalities of management. This must be based on an accurate site map, description and photographs, noting the placement of on-the-spot markers, notices, on-site interpretive information, fencing and possible hazards to visitors. The plan should specify the objectives of management, and provide for a stated system of monitoring that determines whether or not the objectives are being met.

Some monuments, e.g. plaques and memorials, may require no practical management beyond periodic checking to see that they still exist and are intact. Other more vulnerable sites, e.g. graves, cairns and stone huts, may need unobtrusive markers that help guides to keep visitors at a safe distance, with periodic refurbishing and checking to see that the markers remain in place and the fabric of the monument intact. Others again, notably expedition huts with historically interesting contents, need more elaborate management plans, including provision for resident or itinerant caretakers, numbers and conduct of visitors allowed in, and where necessary, delegation of responsibility to nominees in charge of parties when no caretakers are on site. There must also be provision for closing or completely restricting access to any site.

The basic problem in Antarctic historic site mangement – the basic difference between management in the Arctic and Antarctic – is the lack of income, derived from tourist visits, that can be used for mainenance and management of sites. Again Hughes and Davis comment:

> Although tourist visits are levied at some sites by a charge per passenger on cruise ships (Antarctic Heritage Trust, 1993), this is more to reimburse administrative costs, license fees or the distribution of visitor codes, than to meet major logistic and conservation costs. Indeed, some Treaty nations claim that although they would like to ensure the preservation of these sites, many other needs and priorities apply... A "user pays" approach may increasingly be employed for tourist visits but it will constitute only a small element of fiscal needs in heritage conservation which requires more capital equipment and labour than for most other visitor sites in Antarctica. (Hughes & Davis, 1995: 248)

Quite clearly, fees collected from the few dozen or few hundred annual visitors to remote huts cannot be expected to meet the costs of maintaining those huts in perpetuity. But an overall charge *per capita* on the tens of thousands of tourists who now visit Antarctica annually might go far toward paying for some of these and other amenities that they and the tour operators at present receive without charge.

Summary and Conclusions

Arctic culture and heritage are based largely on human populations that have inhabited the Arctic continuously for over 20,000 years. The original inhabitants, nomadic hunters and gatherers who lived along the coasts and rivers, owned very little and left few traces, but several successive cultures have been identified. They became subject to entrepreneurs from the south, who exploited seals and whales for pelts and oil, and traded for furs from land mammals, eventually to be replaced by administrators and other settlers whose imposed cultures culminated in the modern settlements and towns.

Currently, only about 10% of the Arctic population is made up of indigenous people, but it is their cultures and heritage that tourists are most interested to see. While heritage and historic sites were in the care of southern governments, associated tourism raised many problems for the indigenous folk. Recent trends toward self-government have encouraged local communities toward more successful tourist enterprises, which now strengthen local economies and indigenous cultures; though problems remain where tourism grows faster than communities can cope.

Antarctica has no natives and no native cultures or heritage. Historic attractions reside in its many historic sites and monuments, including expedition huts, whaling station sites, graves, memorials and place names. Management principles for heritage and historic tourism are outlined; we draw attention to the lack of funds available for maintaining recognised heritage sites in a region where lack of sovereignty entails lack of revenues from tourism.

Chapter 8
Southern Oceans and Antarctic Tourism

Introduction: Regulation

Islands of the southern oceans are becoming increasingly popular as venues for tourism. In a survey of how tourism is developing in this region, Tracey (2007: 265) lists 23 such islands or groups, spanning a wide range of latitudes and environmental zones from warm temperate to Antarctic. Of these, five lie within the Antarctic Treaty area (i.e. south of latitude 60°S), in which tourist visits are regulated under the Antarctic Treaty System (ATS). Islands north of 60°S are claimed and administered by sovereign governments, each of which provides its own regulations covering tourism. This chapter discusses regulation of tourism (a) on a selection of southern oceanic islands under sovereign governments, and (b) on Antarctica and islands within the Antarctic Treaty area.

The Oceanic Islands

Southern oceanic islands and groups range in area from 1.4 km^2 (Bounty Islands, south of New Zealand) to 13,000 km^2 (Falkland Islands). Their climates are characterised by frequent depressions, resulting in strong, predominantly west winds. Those that rise above a few hundred metres carry a capping of snow in winter. Those that are mountainous (notably South Georgia and Heard Island) have permanent icecaps and glaciers down to sea level, with evidence of more extensive glaciation in the recent past.

The islands lie within four climatic and vegetation zones (see Figure 8.1):

- *Warm temperate islands* include Tristan da Cunha, Gough Island, Ile St Paul, Ile Amsterdam, Auckland Islands and Snares Islands. Lying north of the 10°C isotherm for the warmest month, these are warmer, sunnier and drier than their southern counterparts, with mean summer temperatures 14–17°C, mean winter temperatures 9–12°C. Fertile soils support tussock grass meadows, convertible for pasturing and crop cultivation, with shrub forests where there is shelter from the strong prevailing winds.
- *Cool temperate islands* include the Falkland Islands, Iles Kerguelen, Iles Crozet, Marion Island, Prince Edward Islands, Macquarie Island, Antipodes Islands and Campbell Island. Lying close to or north of the Antarctic Convergence, these islands have equitable

Figure 8.1 Positions of the southern oceanic islands and groups.

maritime climates with mean summer temperatures 6–9°C, mean winter temperatures 2–3°C. Peaty soils support tussock grasses and other flowering plants: extensive moorlands on some of the islands have been converted to pasture for sheep and cattle.

- *Sub-Antarctic islands* include South Georgia, Heard Island and the McDonald Islands. Free of pack ice throughout the year, they have markedly maritime climates with cool summers (mean temperature of the warmest months 0° to + 6°C) and only slightly cooler winters. At sea level, humic, peaty soils support mosses, ferns and tussock grass meadows, grading to moorland and fellfield inland. Soils are thick enough to accommodate colonies of burrowing petrels. South Georgia's moorlands and coastal flats support herds of introduced feral reindeer.

- *Antarctic islands* include the South Orkney, South Shetland and South Sandwich islands. Lying south of the northern limit of pack ice, these have cool summers and hard winters (mean temperature of the warmest months -1° to $+2^\circ$C, winter means down to -15°C). Poor, ahumic soils support meagre floras of mosses, algae, lichens and only two species of flowering plants.

All the islands were exploited by 19th-century sealers, and a few (South Georgia, Iles Kerguelen, Auckland Islands) by whalers. Both left traces of their industries that are now regarded as heritage sites to be protected. Sealers brought few materials ashore and left little more than clay pipes, scraps of canvas, graves and try-pots – the cast-iron pots in which they rendered seal blubber into oil. Whalers left whole whaling stations with boilers, engineering shops, dry docks, scuttled ships and graveyards, all now more or less derelict and regarded as industrial archaeology. They also left legacies of introduced mammals, including house-mice, rats, goats, sheep, pigs and reindeer, which have ensured that few of the islands retain their pre-19th-century flora or fauna.

The main current interest of these islands to their claimant governments is the 200 mile-wide Economic Exclusion Zone (EEZ) surrounding them, in which ownership confers exclusive fishing rights that can bring revenues. Tourism is becoming economically significant to the populated islands (Tristan da Cunha and the Falklands) and to South Georgia. As yet, it brings meagre returns to governments managing the more remote islands, none of which seems likely to encourage growth of tourism for purposes of increasing income.

Attractions to tourists are the islands' remoteness, historic interest, wilderness qualities and wealth of marine-based wildlife, including huge colonies of breeding seabirds and seals. Albatrosses, giant petrels and flocks of smaller seabirds follow the ships as they approach land; penguins, fur seals and elephant seals abound on the beaches. A strong counter-attraction is the islands' distance from gateway ports of the southern continents, across some of the world's roughest seas. The islands are usually visited by cruise ships *en route* to or from Antarctica, requiring a diversion that adds cost to already expensive cruises and can prove uncomfortable in rough weather. Some groups are poorly charted and difficult to approach, requiring landings through surf on exposed beaches.

All but one of the governments that claim southern islands also claim parts of Antarctica (Table 8.1): South Africa alone has no mainland claim, though the South African National Antarctic Programme has maintained a series of research stations on Dronning Maud Land, immediately to its south. All have published measures for environmental protection of their islands and surrounding EEZs, including ordinances for dealing with tourism, which are outlined and summarised below.

Table 8.1 Claims of eight nations to sectors of Antarctica and/or southern oceanic islands

State	*Oceanic islands north of 60°S*	*Antarctic south of 60°S*
Argentina	South Georgia, South Sandwich Island	Antártida Argentina including South Shetland Island and South Orkney Island
Australia	Macquarie Island, Heard and McDonald Islands	Australian Antarctic Territory
Chile		Territorio Chileno Antártico, including South Shetland Island
France	Iles Kerguelen, Iles Crozet, Ile Amsterdam, Ile St Paul	Terre Adélie
New Zealand	Snares Island, Bounty Island, Antipodes Island, Auckland Island, Campbell Island	Ross Dependency
Norway	Bouvet Island	Dronning Maud Land, Peter 1st Island
South Africa	Marion Island and Prince Edward Island	
United Kingdom	Tristan da Cunah, Gough Island, Falkland Islands, South Georgia, South Sandwich Islands	British Antarctic Territory including South Shetland and South Orkney Islands

Note: The Falkland Islands, South Georgia and the South Sandwich Islands are claimed by both Argentina and the UK, but are currently under UK administration. Antarctic territories claimed by Chile and Argentina are deemed to be continuous with their respective mainlands.

Tristan da Cunha and Gough Island

Tristan da Cunha, Nightingale and Inaccessible islands form a closely grouped triangle in the southern Atlantic Ocean midway between South America and southern Africa. Together with Gough Island, 350 km away to the southeast, they are styled the British Overseas Territory of Tristan da Cunha, administered from St Helena. They can only be reached by sea, but have hitherto been rarely visited except by ships concerned with the main industry of cray-fishing. Cruise ship visits are increasing, and developing tourism is included in the island government's management plans.

Tourists are welcome by prior arrangement at the single township of Edinburgh, on the main island of Tristan da Cunha. Landing in the small harbour is only possible in calm weather. Operators are charged for use of the port facilities, and a landing fee is payable for every visitor. The main attraction is the settlement itself, spread over a narrow green shelf overshadowed by the single dramatic peak. The community of 300–350 people welcome strangers guardedly, concerned to avoid the coughs and colds that frequently follow a visit. They provide handicrafts for sale, notably, spun wool, knitwear and scale models of their traditional boats, and support a small museum of community history.

Diversions during a day's stay ashore are walks over coastal pastures to the 'potato patches', where islanders grow their vegetables. More dramatic is a stroll across the lava flow that emerged close to Edinburgh in 1961, resulting in the evacuation of the island population for two years. If time allows, guided parties may climb to at least the lower levels of the main peak. Visitors accompanied by local guides may also land on neighbouring Nightingale Island to see colonies of rockhopper penguins, nesting albatrosses and confiding land birds. No visits are allowed to Inaccessible or Gough Islands, both of which are nature reserves and UNESCO World Heritage Sites.

The Falkland Islands

Lying 600 km (370 miles) east of southern South America, the Falklands form a compact group of two large islands (East and West Falkland) and some 400 small islands, with a total land area of about 12,000 km^2 (4690 sq miles) and a population of c. 2500, of whom over half live in the single town of Stanley. For an account of their geography, history, natural history and recent development, see Stonehouse (2006: 67–82). For an illustrated tourist guide, see Summers (2005).

The Falklands have recently acquired prosperity in which tourism is increasingly prominent. Invasion by Argentine forces in 1982 and subsequent liberation by British forces was followed by an injection of capital from the UK, which provided, among other benefits, a military airfield capable of landing jet passenger aircraft, and facilities for managing fishing within the EEZ. Licensing fishing to foreign ships is currently the islands' main source of income.

The airfield and enhanced port facilities have allowed for the development of local tourism, plus an enhanced share of Antarctic tourism. Airborne passengers spend vacations in Stanley and the 'camp' (hinterland and outlying islands), and also join or leave Antarctic cruise ships. The town meets the needs even of large cruise ships *en route* to and from Antarctica: Bertram *et al.* (2007) style it a gateway port along with Ushuaia, Punta Arenas, Cape Town, Hobart and Christchurch/Lyttelton.

While the ships refuel and restock, Stanley provides souvenir shops, cafés, restaurants and a fleet of taxis and coaches that run full-day or half-day guided excursions to penguin colonies, battle grounds and other local attractions. On a busy day, Stanley may handle 2000–3000 visitors – considerably more than its own population.

Smaller ships may tour some of the smaller islands, where the attractions are long walks, abundant natural history (including five species of penguins and large colonies of albatrosses, elephant seals and sea lions), and farmhouse teas with home-made cakes. Owners of the islands benefit more from landing fees and sales than from sheep farming. The Falkland Islands' government actively encourages all current forms of tourism, working to management plans that seek to increase revenues while maintaining environmental integrity.

South Georgia and the South Sandwich Islands

Administered jointly from the Falkland Islands, these constitute the British Overseas Territory of South Georgia and the South Sandwich Islands. South Georgia, a heavily glaciated island, 200 km long, on the northern arm of the Scotia Arc, is often included in the itineraries of cruise ships *en route* to or from Antarctic Peninsula. With many sheltered bays, harbours and beaches, such historic attractions as an abandoned whaling station and museum, and easily accessible beaches crowded with king penguins, it grows in popularity year by year. For an illustrated visitor's guide, see Poncet and Crosbie (2005). The South Sandwich Islands, a chain of smaller islands at the eastern end of the Arc, lie further from popular routes, have no harbours and few landing points, and are in consequence seldom visited.

Though South Georgia was visited from time to time by cruise ships before and during the early 1990s, the government took little interest in developing tourism until the end of the century, when it perceived opportunities both for regulating a growing industry and providing revenues. Snyder and Stonehouse (2007: 251) report on subsequent growth of the industry:

- in 1991/1992, 11 ships brought 954 visitors;
- in 2000/2001, 27 ships brought 3873 visitors;
- in 2005/2006, 49 ships brought over 9400 visitors.

A comprehensive management plan (McIntosh & Walton, 2002), including objectives and regulations for growing tourism, is reviewed on a five-year cycle. Cruise ship operators require permits to land at a limited range of specified sites. Permitting enables regulation both of types of vessels visiting and numbers of visitors operating at specified sites. Operators that are members of IAATO are allowed a wider selection of sites than non-members. Landings are prohibited in well-publicised scientific reserves.

Landing fees are charged. Tracey (2007: 281) concludes that among all the southern oceanic islands, only South Georgia appears 'to be making money on a scale that bears some relation to management effort and costs'. Considering the island's many resources, Snyder and Stonehouse (2007: 257–260) have outlined a possible multiple resource management approach for further tourism development which is discussed further in Chapter 9.

French oceanic islands

Iles Crozet and Kerguelen, Ile Amsterdam and Ile St Paul are four island groups in temperate latitudes of the Indian Ocean. The first two are extensive archipelagos of steep rugged islands in the cold temperate zone, covered in tussock grass, moorland and shrub vegetation. In the early 20th century, Iles Kerguelen provided a base for whaling, commercial fishing and sheep farming. Ile Amsterdam, a single warm temperate island with similar features, is much modified by fire and earlier attempts at pasturing for sheep and cattle. Ile St Paul, also warm temperate, is a tiny crater island with almost vertical tussock-covered walls, and a small shingle-beach with the remains of a 1930s' cray-fish processing factory.

All are administered by Terre Australes et Antarctiques Françaises (TAAF), a government agency based in Ile Réunion, which operates meteorological and scientific research stations on Iles Crozet, Kerguelen and Amsterdam. These are serviced by a passenger-cargo and research ship, *Marion Dufresne*, which makes three or four voyages each year from Réunion. The ship carries up to 15 tourist passengers, who embark for voyages of 28 days. At sea, they receive instruction on the work of TAAF. Ashore on the islands, they visit the research stations and take part in guided excursions to see penguins, seals and nesting petrels.

None of the islands has an airstrip, and all are too remote from shipping lanes to attract many cruise ships, apart from adventurers in ocean-going yachts. Nevertheless, tourism on each island is subject to stringent conditions of access (available at the TAAF website, www.taaf.fr), which specify preliminary notification of intent to visit, where landings may and may not be made, and regulations for behaviour ashore.

South African and Australian southern islands

These widely scattered islands and groups of the southern Indian and Pacific oceans share a common heritage of 19th-century sealing. Largely unclaimed until the early 20th century, they have since been annexed by their closest neighbouring states, which benefit principally from the fishing rights that can be claimed within 200 mile-radius EEZs.

South Africa manages Marion Island and Prince Edward Island – two islands lying close together in the Indian Ocean, some 1800 km southeast of Cape Town. Cool temperate islands, ringed by steep cliffs with few points of access, tussock-covered and rich in wildlife, they are administered as special nature reserves of Cape of Good Hope Province. A scientific research station operates on Marion, the larger island, as part of the South African National Antarctic Programme (SANAP: for details see http//marion.sanap.org.za). The islands' potential for tourism has been assessed as part of a management plan. Controlled tourist visits to Marion Island may become possible in the future, but Prince Edward Island is likely to remain a reserve for scientific research.

Australia manages Heard Island and the McDonald Islands, which lie some 4000 km from the mainland in the sub-Antarctic zone. Heard Island has a single glaciated mountain with surrounding strand flats and beaches, and several volcanic vents that have erupted mildly in recent decades. The three McDonald Islands, 60 km to the east, are smaller and low lying, recently more than doubled in area by new lava flows. Despite visits from sealers extending into the 20th century, this group is free from introduced mammals and has few alien plants.

With their surrounding EEZ, these islands form a scheduled Marine Reserve, administered from Hobart by the Australian Antarctic Division of the Department of the Environment and Heritage, and listed also as an IUCN World Heritage Site. Extreme remoteness and difficulties of landing protect them from casual visits. Under a management plan issued in 2005 (see http//www.heardisland.aq), visits to Heard Island are allowed under permit, with restrictions on numbers allowed ashore and strict quarantine regulations to keep out undesirable plants and animals. Landings on the McDonald Islands are prohibited except in an emergency.

Australia also manages Macquarie Island, a cold temperate island on the Antarctic Convergence south of Tasmania. Tussock-covered and rich in wildlife, it is designated a national park within the Tasmanian Parks and Wildlife Service, and an IUCN World Heritage Site. A scientific station is managed by the Australian Antarctic Division. The island is visited mainly by Antarctic cruise ships. Visitors are limited to a few hundred per year, who are met ashore by park rangers and provided with guided tours of the settlement and environs.

New Zealand's southern islands

New Zealand manages five groups of islands lying to the south and east of the South Island – the warm temperate Snares Islands, Auckland Islands and Bounty Islands, and cool temperate Campbell Island and Antipodes Islands. Their flora and fauna are described in UNEP-WCMC (2006); see also www.unep-wcmc.org/sites/wh/subanta.htm. Fraser

(1986) covers their history and wildlife. Of their ecological standing, Tracey writes:

> It would be difficult to overstate the significance of the wildlife populations of these islands. Breeding birds number in the millions – the Snares Islands alone host an estimated six million birds. Globally significant populations of many endemic, rare and endangered species are present, including the world's rarest cormorant, duck, snipe and penguin species. Marine mammals include the rare and endangered New Zealand sea lion, elephant seals and New Zealand fur seals. (Tracey, 2007: 268)

Some are no less important floristically, though Campbell Island was severely modified by sheep farming and Enderby Island, in the Auckland Islands group, supported a short-lived 19th-century whaling settlement, traces of which can still be found in the coastal forest (Dingwall *et al.*, 1999). All are scheduled nature reserves. The Auckland Islands and Campbell Island may be visited; operators must submit an environmental assessment that, among other points, shows an educational component to their visit, and numbers of visitors are restricted. Campbell Island offers seals, penguins, albatrosses and tussock meadows. On the Auckland Islands, visitors meet sea lions and roam through forests of flowering shrubs. No landings are permitted on the remaining three groups.

Summary: Southern islands management

The authorities that manage the southern islands agree substantially on what is required for effective tourism management. Tracey (2007: 270–274) tabulates accord between regulations for the New Zealand and Australian islands, Tristan da Cunha and South Georgia, and compares them with those for other southern oceanic islands. Bertram and Stonehouse (2007: 287–289) make similar comparisons between regimens for South Georgia and two Arctic localities – Svalbard and Glacier Bay (Alaska) – comparing the regulations and constraints imposed at these venues with those applying to Antarctic Peninsula under the ATS.

Tracey (2007: 280–281) lists the following key points that appear common to all regulations covering the southern oceanic islands:

- Management includes planning both for conservation and for controlled tourism.
- Management is conservative, precautionary and protective, taking account of the islands' important natural values and reserve status.
- Particular objectives are to avoid introducing alien species and disturbance to wildlife by controlling access, requiring quarantine

procedures, selecting sites where tourism can be conducted without significant impacts and closing islands where risks are high.

- Most authorities recognise the educational benefits of tightly controlled tourism to some islands, and closing others to casual or recreational visitors.
- Visits are allowed only to sites where safe landing is possible, or where good viewing may be achieved with minimal environmental impacts.
- Limitations on numbers of visits or visitors per season tend to be arbitrary and low – approaches that reflect different conditions, different levels of tourism use and different levels of precaution adopted by managing authorities.
- The islands remain relatively undamaged by tourism, sometimes after many years – a tribute to the adequacy of management systems and the ecological integrity of small, properly managed shipborne operations.
- Managers prescribe regulations, but also accept the need for self-regulation of tourist activities through codes of conduct, and advise on appropriate behaviour and ways in which requirements may be met.

Tracey's final conclusion is that:

> The management systems covering all these islands show marked similarities, and are to some degree based on each other. ...the overall similarity of management provisions suggests that a model of "best-practice" management has emerged for these areas – a model that is worthy of consideration in many remote wilderness areas and locations subject to increasing pressures of tourism.

Bertram and Stonehouse (2007) reach similar conclusions in their comparisons of management at Arctic tourist venues with those under the ATS.

Regulation in the Antarctic Treaty Area

The Antarctic Treaty arose in 1959 in the aftermath of the International Geophysical Year (IGY) of 1957–1958. During preparations for the Year's studies, 12 nations opted to undertake geophysical studies in Antarctica, including the seven nations that claim sectors of the continent (Chapter 1) and five others (Belgium, Japan, South Africa, the Soviet Union and the USA). Faced with the certainty that at least the USA and USSR would ignore claims and work where they saw fit, the claimant nations agreed that scientists of all the IGY nations could work freely within their sectors.

As much of the research was set to continue after the end of the IGY, in some 40 stations that were already established all over the continent, the agreement to allow free access also continued. In 1958, the US government proposed to the 11 other states that a treaty be negotiated to continue scientific cooperation and ensure that Antarctica would continue to be used for peaceful purposes. The Antarctic Treaty was signed by representatives of the 12 nations in 1959, and came into force on 23 June 1961. For details of the Treaty's origins, early development and legal implications, see individual papers in Triggs (1987).

The Antarctic Treaty system

The Antarctic Treaty came into being at the height of the Cold War, at a time when the USA and USSR on the one hand, and Argentina, the UK and Chile on the other, were deeply suspicious of each others' motives in Antarctica and elsewhere. It gave diplomatic representatives of the 12 IGY nations opportunities to negotiate agreements on a wide range of topics, in the interests of keeping Antarctica from becoming 'the scene or object of international discord'. For the Treaty's main provisions, see Appendix E.

Antarctic tourism began only a few years before the Treaty was negotiated. Not surprisingly, neither young organisation took note of the other. If delegates to early Antarctic Treaty Consultative Meetings (ATCMs) (then held at two-yearly intervals) had misgivings about the activities of cruise ships, they were not recorded. Among topics specified by the Treaty as appropriate for discussion were 'measures regarding... the preservation and conservation of living resources in Antarctica'. This was one on which Consultative Parties felt that agreement could readily be reached by the necessary consensus, so environmental protection became one of the earliest issues discussed (Heap: personal communication to Stonehouse, 1994). By 1964, within three years of the Treaty's coming into force, representatives had negotiated 'Agreed Measures for the Conservation of Antarctic Flora and Fauna', which made no mention of tourism, but required environmentally responsible behaviour from all who visited the continent.

Tourism was first noted in the 4th ATCM of 1966, which recognised that 'the effects of tourist activities may prejudice the conduct of scientific research, conservation of flora and fauna and the operation of Antarctic stations'. As the industry expanded, the misgivings of the scientific community grew. The Treaty's stated intention that Antarctica be 'a continent for peace and science' could readily be interpreted as excluding other activities that had no bearing on those objectives; thus tourism, being irrelevant to either, had no place in the Antarctic Treaty area. This view was firmly negated at the 8th ATCM of 1975, which acknowledged that tourism was 'a natural development in the Area', adding that it required regulation. The same meeting recommended the designation of 'Areas of Special Tourist Interest' that tour operators would be requested to use exclusively. No such areas were ever defined: no reasons were given for the default, but it was clearly a matter on which the necessary consensus was unlikely to be reached.

The Treaty parties gradually assumed management responsibilities in a wide range of fields. In 1972 came the 'Convention for the Conservation of Antarctic Seals', and in 1980 the more ambitious and far-reaching 'Convention for the Conservation of Antarctic Marine Living Resources' (CCAMLR) – an attempt to control the growing fisheries industry of the Southern Ocean. These developments provoked growing concern among world environmental movements that the Treaty nations, meeting in conclave from which non-members and press were excluded, were allowing Antarctica and its surrounding ocean to be ruined by commercial

Box 8.1 Antarctica a World Park?

The Second World Conference on National Parks, meeting in the USA in 1972, recommended that the Antarctic Treaty nations should establish the continent and surrounding oceans as the first World Park, under the control of the United Nations. In 1981, the General Assembly of the International Union for Conservation of Nature and Natural Resources (IUCN), recalling the 1972 proposal, urged the Treaty parties to 'further enhance the status of the Antarctica environment' and ascribe to it 'a designation which connotes world-wide its unique character and values'.

The call was renewed in the following year at a session of the United Nations Environment Programme (UNEP), which requested the Treaty powers and the UN General Assembly to consider declaring Antarctica a World Park. Like the earlier appeals for a World Park or similar designation, this call failed to indicate in more than very general terms what was intended.

The Antarctic and Southern Ocean Coalition (ASOC), representing a strongly conservationist view for the environmental protection of Antarctica, also supported the World Park concept. ASOC's more pressing concerns at that time, expressed in a popular book, *Let's save Antarctica* (Barnes, 1982), were the danger from commercial fishing, already in operation, and from mineral extraction that appeared imminent. The Treaty's CCAMLR, recently announced, was not in their view adequate to control harvesting of fish stocks. They were equally concerned that secret negotiations were in train to permit minerals exploitation on the continent (the ultimately discredited CRAMRA: see below) with results that were likely to be devastating to the environment. Neither they nor anyone else convinced the Treaty powers of the merits of a World Park.

exploitation – by fisheries through CCAMLR, and potentially by mining operations, on which a convention was under consideration. From these misgivings arose a movement to take Antarctica out of the exclusivity of ATS stewardship and declare it an 'International Park' or 'World Park'.

The Treaty nations were indeed at this time developing a convention to regulate mining on the continent. A succession of special meetings during the period 1982–1988 established a Convention on the Regulation of Antarctic Mineral Resources (CRAMRA), as an additional instrument in what was coming to be known as the Antarctic Treaty System (ATS). CRAMRA was introduced at a special meeting in Wellington, New Zealand, in June 1988. It was subsequently rejected by the governments of two nations, France and Australia, both responding to strong pressures from their own national environmental lobbies, on grounds that it gave insufficient protection to the Antarctic environment. As consensus was required, CRAMRA never came into force. In response to strong reaction from conservation bodies and the public, mining activities were instead banned under a 50-year moratorium.

Alerted by the failure of CRAMRA, the Treaty powers addressed themselves more closely to environmental conservation measures *per se*. By 1991, they had elaborated a new instrument, the 'Protocol on Environmental Protection to the Antarctic Treaty', which attempted to remedy some of the weaknesses and inconsistencies perceived in CRAMRA. Introduced at a special meeting in Madrid in 1991, this became known as the 'Madrid' or '1991' convention. For an account of the transition, see Heap (1994: 2001). For a summary of how the Treaty powers sought to regulate tourism at the time when the Protocol came into being, see Enzenbacher (1995). The terms of the Protocol that relate to tourism are discussed in '**The environmental protocol**' (p. 155).

The development of the International Association of Antarctica Tour Operators

Anticipating the possibility of severe restrictions being placed on tourism by the Treaty powers, in 1991 seven corporate members of the industry formed IAATO. Developed on the initiative of Lars-Eric Lindblad, IAATO was intended 'to act as a single voice in concerns of tourism and to advocate, promote and practise environmentally responsible private-sector travel to Antarctica' (Landau & Splettstoesser, 2007: 198). IAATO quickly made its mark with sets of guidelines, addressed, respectively, to Antarctic visitors and to staff and crews of tour operators, which were well in advance of any form of regulation thus far produced by the ATS, and later formed the basis of Treaty-based legislation.

IAATO was quickly recognised by the Treaty powers as a responsible representative of an industry that they were unable to control directly. From 1994, the Association participated with observer status in ATCMs. A part-time secretariat initiated a website (www.iaato.org) for the benefit of members and the interest of the public, which has since kept accurate records of its members' activities, and published annually the names of ships and operators, numbers of cruises and passengers, landing sites used and other details, which are of considerable importance in tracing the growth of the industry.

IAATO has maintained its prominence in the industry: for an outline of its current rôle and standing, see Landau and Splettstoesser (2007: 198–203), who list 18 highlights that the organisation achieved since its inception. These include:

- successful management of Antarctic tourism by establishing mechanisms for operators to work together, though all are competitors of each other;
- development of ship-scheduling programmes to coordinate visits to landing sites;
- development of an online database of information to assist in emergency contingency planning;
- compilation of tourism statistics;
- compilation and issuing of Site Guidelines based on 12 years' information from field staff, to establish time limits and numbers of passengers for environmentally safe visits, and development of site selection criteria, activity guidelines and operating procedures to ensure compliance of tourism standards;
- introducing boot washing and clothing decontamination to prevent alien species from contaminating Antarctica;
- improvement of navigational charts by submitting newly found data to appropriate charting authorities;
- cooperation with the ATS over development of regulations for tourism;
- support for a scheme for developing inventories of species and other data at landing sites.

This listing includes many items that, elsewhere in the world, would unequivocally have been the responsibilities of the government under sovereignty, but were undertaken by IAATO in its absence – tasks that the ATS had not so far empowered itself to undertake, and at that stage would no doubt have had great difficulty in assuming. Landau and Splettstoesser bring their account of the organisation up to date:

> By 2006 IAATO included some 80 members operating in 14 countries... Most members are ship operators, working as business competitors. But members also include two land-based operators, one

> fly/cruise operator, one company that conducts scenic overflights without making landings, several travel companies who charter ships, one helicopter operator, agencies that cater for adventure travel, shipping agencies, government tourism offices and conservation groups. (Landau & Splettstoesser, 2007: 198)

There can be little doubt that IAATO has compensated for many of the limitations of the Treaty system in regulating tourism south of 60°S. However, that tour operators in the association are filling rôles that are elsewhere (and more properly) undertaken by governments is a matter for comment. As noted by Wouters (1993: 77, cited by Johnston & Hall, 1995: 304):

> The present spirit of cooperation among the major tour operators should be encouraged, but may need to be supplemented by more formal measures which also provide an enforcement mechanism.

This observation, written 17 years ago, is no less true today. 'More formal measures' in the shape of an Environmental Protocol to the Antarctic Treaty (see below) are indeed now in place, but the ATS still lacks many of the powers that were surrendered with sovereignty and have not been replaced, including means of enforcing its measures to regulate tourism.

The environmental protocol

Of several successive sets of environmental measures devised over the years since the 1964 Agreed Measures (p. 151), the 1991 Protocol on environmental Protection to the Antarctic Treaty is the most recent, and most relevant to current tourism management. Presented at the 11th Antarctic Treaty Special Consultative Meeting in Madrid in October 1991, the Protocol was quickly ratified by all the key governments, and has since formed the basic text for regulating all human activities in Antarctica, including tourism.

The full text of the Protocol (available at http://www.antarctica.ac.uk/about_antarctica/geopolitical/treaty/update_1991.php) begins with a preamble and a series of definitions. In Article 2, the Parties commit themselves to the comprehensive protection of the Antarctic environment and designate Antarctica a 'natural reserve, devoted to peace and science' – without, however, assigning any precise meaning to the phrase 'natural reserve'. In Article 3, headed 'Environmental principles', Paragraph 1 specifies that:

> The protection of the Antarctic environment and dependent and associated ecosystems and the intrinsic values of Antarctica, including its wilderness and aesthetic values and its value as an area for the

> conduct of scientific research, in particular research essential to understanding the global environment, shall be fundamental considerations in the planning and conduct of all activities in the Antarctic Treaty area.

We have already discussed possible interpretations of 'wilderness and aesthetic values' in Chapter 5 (p. 82). The statement 'value as an area for the conduct of scientific research' is less ambiguous, and reinforced by Paragraph 3 of the same section:

> Activities shall be planned and conducted in the Antarctic Treaty area so as to accord priority to scientific research and to preserve the value of Antarctica as an area for the conduct of such research, including research essential to understanding the global environment.

Having reiterated the pre-eminence of science in Antarctic affairs, Paragraph 4 includes a rare mention of tourism:

> Activities undertaken in the Antarctic Treaty area pursuant to scientific research programmes, tourism and all other governmental and non-governmental activities in the Antarctic Treaty area for which advance notice is required in accordance with Article VII (5) of the Antarctic Treaty, including associated logistic support activities, shall:
>
> (a) take place in a manner consistent with the principles in this Article; and
> (b) be modified, suspended or cancelled if they result in or threaten to result in impacts upon the Antarctic environment or dependent or associated ecosystems inconsistent with those principles.

Article 3(2) and ensuing articles require all activities in the Treaty area to be planned and conducted in ways that limit adverse impacts on the environment, do not increase the jeopardy of endangered or threatened species, or hazard areas of biological, scientific, historic, aesthetic or wilderness significance. They must be planned and conducted on the basis of information sufficient to allow prior assessment of, and informed judgements about, their possible environmental impacts, and must provide for regular and effective monitoring to ensure that no adverse changes are occurring. They must accord priority to scientific research, preserve the value of Antarctica as an area for the conduct of research, and be modified, suspended or cancelled if they prove detrimental.

Articles 11 and 12 of the Protocol provide for the establishment of a Committee for Environmental Protection (CEP), the functions of which are:

> to provide advice and formulate recommendations to the Parties in connection with the implementation of this Protocol, including the operation of its Annexes, for consideration at Antarctic Treaty

> Consultative Meetings, and to perform such other functions as may be referred to it by the Antarctic Treaty Consultative Meetings.

First meeting in 1998, the Committee has provided annual reports to ATCMs on environmental matters within its competence. Reports are available on the ATS website (www.ats.aq/e/cep.htm).

As a statement of the impeccable intentions toward the Antarctic environment of over 40 nations in conclave, the Protocol would be difficult to improve on. As a basis, among other considerations, for managing so ebullient an industry as tourism, it is arguably less than adequate. Much detail that is important to practical management was left out, to be decided by later interpretation. This is entirely understandable in the context of diplomacy. Had the representatives attempted to incorporate detail, the Protocol would no doubt still be unadopted and under discussion at Consultative Meetings.

However, though much has been discussed and written on interpretations of the Protocol since it came into force, and progress has been made toward recognition of some of the environmental problems arising from tourism, effective regulation of the industry, along lines that are generally accepted as applicable to similar areas elsewhere in the world, has not so far appeared.

Environmental impact assessment

In practice, for permitting any human activity within the area south of 60°S, the Protocol relies on an environmental impact assessment (EIA) procedure, stated in Annex 1, which Hemmings and Roura (2003: 21) identify as 'the sole gatekeepers for Antarctic access'. The procedure is outlined in Figure 8.2.

Under Article 1 of the Annex, the environmental impacts of proposed activities:

> shall, before their commencement, be considered in accordance with appropriate national procedures... If an activity is determined to have less than a minor or transitory impact, the activity may proceed forthwith.

Neither 'appropriate national procedures' nor 'minor or transitory impact' are defined, so interpretation is left to the judgement of individual governments. As Kriwoken and Rootes (2000: 142) point out, different states have different legal systems, providing variations in how ambiguities and uncertainties in the wording are interpreted. Different states also have different standards and criteria for what is acceptable in tourism, different concepts of where the limits to tourism lie, and different models for the practicalities of management that the Protocol lacks.

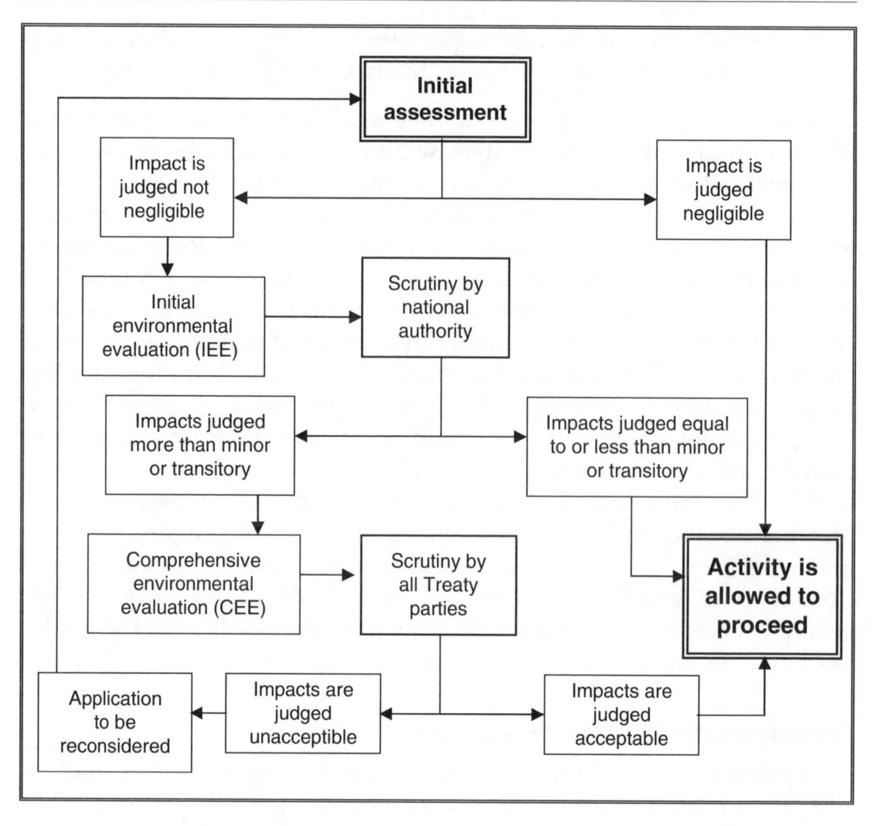

Figure 8.2 Schematic diagram of environmental impact assessment under Annex 1 of the Protocol. After Crosbie (1998: 40) and Bertram (2005: 49)

Unless the proposed activity is judged to have less than a 'minor or transitory impact', Article 2 of the Annex requires an Initial Environmental Evaluation (IEE) for consideration by the home government. The IEE:

1. Shall contain sufficient detail to assess whether a proposed activity may have more than a minor or transitory impact and shall include:
    a. a description of the proposed activity, including its purpose, location, duration and intensity; and
    b. consideration of alternatives to the proposed activity and any impacts that the activity may have, including consideration of cumulative impacts in the light of existing and known planned activities.
2. If an Initial Environmental Evaluation indicates that a proposed activity is likely to have no more than a minor or transitory impact,

the activity may proceed, provided that appropriate procedures, which may include monitoring, are put in place to assess and verify the impact of the activity.

Should the activity be judged to have more than a minor or transitory impact on the environment, a more searching Comprehensive Environmental Evaluation (CEE), fully prescribed in Article 3, is required. A draft CEE must be circulated to all Parties for comment within 90 days, and at least 120 days before the next ATCM.

Discussions at successive ATCMs (Heap, 1994) revealed many points of confusion among delegations, arising from ambiguities and uncertainties within the wording of the Protocol, mostly in relation to the activities of national expeditions and research stations that the Protocol was designed primarily to regulate. Gradually, these appear to have been resolved to the satisfaction of delegates. More fundamental difficulties have arisen when provisions of the Protocol have been applied to tourism, which operates in different ways from national expeditions and presents different environmental challenges.

For example, in 1996 when the requirement for IEEs was first made, IAATO immediately devised 'programmatic' IEEs to cover standard landings, which would make it easier both for tourism operators to apply for permits and for permitting authorities to allow them. This was a pragmatic and sensible way of 'clearing' a difficulty that might otherwise have held up a season's tourist activities, and was clearly acceptable to the governments involved. However, it ignored the more fundamental issue of whether all sites can be judged equally vulnerable to landings (Bertram, 2007: 166). Quite clearly, they are not equal, for they vary greatly in spatial distribution of flora, fauna, historical associations and other attractions, and correspondingly in vulnerability to damage by landing operations.

What might have been regarded as an expedient to cover a single season has not been reconsidered, and programmatic IEEs are still accepted. Since the policy was adopted, numbers of ships visiting and passengers landing have increased substantially (p. 55), yet there is no mechanism to determine whether this particular labour-saving device continues to provide valid protection for landing sites. Thus, all landing activities currently in force continue to be judged as having 'less than a minor or transitory impact'. Do they in fact have almost negligible impact at all the landing sites? The regulators have provided no means of judging.

What is needed?

Thus the Protocol, while providing generally for protection from human intrusions, does not address the issues arising from tourism for which sovereign governments in the Arctic, and throughout the southern oceanic islands, regard as important, and for which they provide in management plans. Bertram and Stonehouse (2007: 285–286) list seven 'critical elements' that the ATC has so far been unable to address, on which the following points are based (Figures 8.3–8.5).

Figure 8.3 A well-behaved visitor photographs Adelie penguins on the icefoot of an Antarctic island. Photo: BS.

Figure 8.4 Moulting elephant seals on an Antarctic beach. Though normally placid, they can all-too-easily be disturbed by noisy or intrusive visitors. Photo: BS.

Figure 8.5 An inquisitive humpack whale investigates visitors in stationary inflatable boats. Photo: BS.

- Tourism in Antarctica, particularly shipborne tourism to coastal sites along Antarctic Peninsula, continues to expand and diversify at accelerating rates. Market forces alone determine how many ships, passengers and aircraft are involved. The ATS appears to have no provision for limiting visits, should limitation prove desirable.
- There is no overall strategy, based on clearly stated goals and objectives, within the ATS specifically to manage or control Antarctic tourism or its operations in the field. Despite the intrusion of a large and growing industry, no tourist management plans have been drawn up either for Antarctica as a whole, or for individual landing sites.
- Antarctic tourist operations are managed under regulations originating from a Protocol that was designed mainly to cover activities of national expeditions at research stations, and can be applied only with difficulty to the quite different activities of tourists at landing sites.
- Tourist landings are permitted under 'programmatic' (i.e. generalised) IEEs (see **Environmental impact assessment**, p. 159) that take no account of individual differences between landing sites. Tour operators are required to 'monitor' the effects of their activities, but have no regulations, guidelines or advice on how to monitor, or – in the absence of management plans and strategies – what to monitor for (pp. 95–96).

- There are no provisions for monitoring by experienced ecologists independently of the industry, no formal mechanism for reporting of sites at which changes due to tourist pressures may have occurred, or for the investigation or remediation of such changes.
- Though Site Guidelines have recently been issued covering a number of heavily visited landing sites, site-specific management objectives have not been defined for monitoring purposes, and provisions remain unclear for monitoring, reporting damage, possible closure to allow natural recovery or remediation to accelerate recovery processes.
- The ATS has no capacity for enforcing regulations relating to tourism, and no source of income to support ranging, inspection, enforcement, monitoring, remediation of damage or any other expenditure in connection with tourism regulation. Responsibility for the actions of individual operators or tourists lies with their respective governments – responsibility that the governments have no practical means of assuming.

Of these points, arguably the most fundamental are (a) lack of an overall policy for tourism to the continent, in particular to the Peninsula area; and (b) lack of income from the industry to fund an authority responsible for ranging, inspection, monitoring, enforcement and other services that are elsewhere deemed essential for tourist management (Figure 8.6).

It is well within the capacity of the Treaty System to provide for the first, and should not be beyond its capacity to provide for the second. The possibility of a special annex to the Protocol covering tourism was

Figure 8.6 Managing tourist numbers and their behaviour. Photo: JMS.

earlier considered and rejected: for a contemporary comment, see Enzenbacher (1995: 185–186). More practically, the idea of a Commission for Tourism based on the successful model of CCAMLR (p. 152) has been suggested by several observers, including Boczic (1988) and the Antarctic and Southern Ocean Coalition (2004). Bertram concluded that developing another level of bureaucracy within the ATS is probably unnecessary:

> General concern remains focussed on degradation of sites visited by tourists, rather than whether tourism management has a cohesive framework. Even if a tourism convention existed, the only active management option available, if a site showed signs of degradation, is placing the site off limits for a period of time. This would require the ATS to commit to site protection through a simple monitoring strategy. (Bertram, 2005: 255–257)

Such action is indeed all that is required to solve immediate problems relating to site management. However, it leaves untouched the more basic problems of controlling or limiting the activities of the industry as a whole. This will inevitably become necessary as the industry continues to grow, and will require careful planning by a dedicated and responsible body within the ATS. In view of the very rapid initiatives of which the industry is capable, and the very slow responses that the ATS finds possible, the sooner such planning begins, the better for the Antarctic environment.

Bertram's ecological point arises from the absence of provision for reporting tourist-induced damage to landing sites. Bertram and Stonehouse (2007: 304–305) have suggested that in practice, the forms of damage most likely to occur on landing sites are (a) reduction in size of breeding bird colonies (which may, however, be due to environmental changes unrelated to tourism: see Patterson *et al.*, 2003) and (b) man-made tracks over gravel or vegetated surfaces. Should damage for which tourism is clearly responsible become apparent, the only practicable remedy will be to restrict access to part or all of the site, if necessary curtailing visits for several seasons to encourage natural restitution.

What is therefore needed for site protection is not ill-defined 'monitoring' (the need for which is anyway currently ignored), but a simple system for reporting signs of damage, inspection, assessment – possibly by a suitably qualified onboard guide – and a recommendation to the next ATCM that visits to the site be restricted. Under Article 3 of the Protocol, such a site could be declared an Antarctic Specially Protected Area, a category designed:

> to protect outstanding environmental, scientific, historic aesthetic or wilderness values, any combination of these values, or ongoing or planned scientific research.

The necessary planned scientific research would consist of a five-year monitoring programme to measure the site's recovery, renewable if necessary for a further five years under the same ruling.

That so little environmental damage has been done by half a century of tourism in Antarctica is a tribute to the way in which the industry has managed its own operations, in particular to the long-term responsibilities undertaken by IAATO in advocating sustainable practices. The continent's 'self-appointed stewards' of the ATS (Enzenbacher, 1995: 180), with neither authority nor means of exerting authority that are implicit in sovereignty, have yet to achieve the levels of control shown by any sovereign government of the surrounding oceanic islands.

As numbers of visitors continue to grow, controlled entirely by market forces and not at all by environmental considerations, it remains to be seen whether IAATO and the Protocol can provide the protection required to ensure sustainable tourism in a sensitive and precious environment.

Summary and Conclusions

Like Antarctica itself, several islands and island groups of the surrounding southern oceans have become tourist attractions that are increasingly being visited by cruise ships. All are, or include, wilderness areas with attractions of remoteness, scenic beauty, historical heritage and spectacular wildlife, similar to those of Antarctica itself. All are governed by sovereign states, which have independently adopted relatively similar patterns of tourism management, involving forward planning with stated objectives, and regulations covering numbers of visits permitted, landing sites, etc. Tourist visits to many of these islands are currently limited by their remoteness from shipping routes, but should demand from the industry increase, the respective governments have means of control already established.

The Falkland Islands and South Georgia, relatively close to the direct routes from South American gateway ports to Antarctica, in particular have provided for controlled development of tourism during the next few years. The Falklands are already established as a tourist venue: South Georgia has a well-planned, forward-looking programme for tourism, with the potential for becoming self-funding and possibly profitable.

Tourism to Antarctica and the Antarctic Treaty area south of 60°S is, by contrast, managed under the terms of the Protocol on Environmental Protection to the Antarctic Treaty. This seeks to manage all forms of human activity under the same rulings, with no special provisions to cover tourism. The complete absence of provision, under the Protocol, for almost all the planning and management measures that sovereign governments of the southern islands deem necessary, is a cause for concern, particularly in view of the steadily increasing numbers of visitors to the Antarctic area. In practice, the main current problem is ensuring that landing sites are managed in ways that ensure their protection – a problem for which simple remedies are available, though not yet in place. Of more lasting concern is the absence of planning for the overall management of the industry, the rapid expansion of which is currently subject only to market forces. Such planning will inevitably be required in the near future: an early start is recommended.

Chapter 9

Managing Polar Tourism: A Way Forward

Introduction: Regulation and Management

Throughout the entire polar world, tourism is exerting impacts on environmental conditions, institutional resources, economic and social well-being, and cultural integrity. Simultaneously, the polar world itself is altering: both changing environments and growing economic development pressures are transforming the settings within which polar tourism operates. Protecting those environments, while sustaining economic development and preserving integrity, is more than a laudatory goal: it is an essential task, to be undertaken by governments and industry in collusion.

In the Arctic, the task falls where tourism occurs alongside mining, oil-drilling and other economic developments, regulated by institutions with scarce infrastructure and finances. In the Antarctic, tourism operates only alongside science, and the regulating institution has no infrastructure or finances. Slow in response, it is well designed to keep scientists in order. Dealing with a proliferating industry, it relies largely on the integrity of the industry to perform the essential tasks of protection and preservation.

As we show in our first three chapters, Arctic and Antarctic regions are similar but different, and their patterns of tourism have evolved in different ways with different emphases. Chapter 4 points out that this evolution has occurred against a background of change. Not only is the physical environment in constant flux, but man's concepts and uses of polar regions are changing too. In Chapter 5, we assert that the regions' main attractions – their wilderness values – remain constant, and must be held constant by sound regulation and management. Though coupled in the Arctic with the presence of indigenous populations (discussed in Chapter 7) and in the Antarctic with exploration and ongoing scientific research (Chapters 3 and 8), the quality of wilderness must at all costs be maintained if tourism is to continue without ruining both the polar regions and itself.

In Chapter 6, we provide evidence of the rapid growth of cruise ship tourism and the very real threats posed by the alarming rise of marine incidents in polar waters. Between 2000 and 2007, 62 serious marine incidents occurred, including five ships that sank and 17 that were grounded. Management implications are enormous: more ships in polar waters represent increasing threats to human safety and environmental quality, and require efficient management responses. Quick response for search and rescue to save human lives, and the timely employment of oil containment equipment, are the most vital measures of management efficiency. Improved communications, expanded infrastructure and the application of new vessel standards will substantially enhance management capabilities.

Thoughtful approaches both to regulation by governments, and to management by operators, are therefore essential. Some polar governments and tour operators combine forces over management techniques that are well suited to polar environments and their wildlife and cultural heritages. Others do not. It is timely to seek wider application of successful techniques, and design and implement new tourist management systems that address current realities in both polar regions. In this final chapter, we summarise and discuss key factors for providing effective regulatory structures and sound management policies.

Diversity and Growth

Polar tourism is a mature, diverse industry, operating in both polar regions and attracting a wide variety of clients. It takes place in diverse ecological and cultural settings, involving diverse operations and jurisdictional authorities. Clients are motivated variously to experience polar environments, wildlife, culture, history, sporting activities and adventure, throughout the entire polar world and during all seasons of the year.

Though shipborne tourism predominates at both ends of the world, it does not eclipse the importance of land-based tourism in the Arctic. Cruise ships operated by companies from outside the polar regions bring the greatest numbers of visitors to both Arctic and Antarctic. But much Arctic tourism on land and in fresh-water venues involves local people and companies, often practicing newly acquired skills and expertise in sharing their heritage with the outside world.

Thus, polar tourism requires regulation and management to cover its wide range of activities in the settings in which they occur. An obvious conclusion? Yes, but one that is often avoided and clearly needs stating. Approaches, for example, that in effect admonish tourists and tour operators to 'behave properly' or 'cause no harm' are inadequate even in well-patrolled areas. They are considerably less effective in areas where infrastructure is weak and rangers are thin on the ground.

Similarly, threats of legal action against infringements are unlikely to take effect where there is nobody with authority to secure evidence and press charges. Simplistic, generalised regulatory and management systems, designed for non-polar environments, give clear evidences that neither regulators nor managers are attuned to the special needs of polar environments.

The approach of the Madrid Protocol to regulating tourism in the Antarctic, as just another human activity to be controlled in the same way as national expeditions, indicates very clearly the Treaty delegates' awareness of the difficulties of managing a continent by consensus – agreement that must be reached among many nations with as many different agendas for all forms of management. It shows less awareness of the practicalities of managing and controlling tourism. Antarctica, no less than any other area that is vulnerable to tourist impacts, rates sound, well-considered tourism regulation, not only for the present, but for the not-too-distant future when visitor numbers may again have doubled and redoubled.

Regulating for growth

Tourism is growing worldwide, seemingly little affected by economic downturns. The mass tourism that developed elsewhere has spread to the polar regions, including remote and unlikely corners. Simple approaches have given way to more elaborate ones – diversification that allows markets to expand is now evident at both ends of the world.

Increasing numbers of tourists is the single most obvious measure of growth, but regulation and management must deal also with tourism's expanding geographic range, increasing diversity of activities and attractions, growing duration of stay or recreation participation and lengthening seasons of use. All these forms of growth require management responses, including the further dispersal of already scarce management and emergency support resources, and the extension – where they exist – of monitoring and enforcement capabilities.

In predicting that polar tourism will continue to grow, we have identified particularly the following contributory causes:

- *Improved physical access.* Hostile environmental conditions, particularly sea ice and severe weather, represented the most substantial barriers to polar travel. Reductions in extent, duration and thickness of Arctic sea ice are now well documented. Associated with climate change are small increases in daily temperatures, warm seasons of greater duration and changes in precipitation frequency. Ships of all types are transporting tourists to increasingly ice-free destinations, and improved weather conditions are enabling more reliable commercial air transport. The combination of more tolerable weather conditions and vastly improved clothing and recreation equipment also improve the personal comfort of tourists. Inhospitable conditions are steadily being overcome by natural events and technological advancements (Figure 9.1).

Figure 9.1 MV *Spirit of Adventure*, a medium-sized liner, limits its passengers to fewer than 500, thereby qualifying under IAATO rules to be able to take passengers ashore in Antarctica. Photo: BS.

- *Growing awareness and popularity.* Arctic tourism has grown in popularity for more than a century, Antarctic tourism for over half a century, through good economic times and bad. With increasingly convenient schedules, more attractive facilities, more comfortable transport and the increasing willingness of the media to advertise already popular venues, there is every reason to believe these trends will continue. Steadily increasing personal income, leisure time, educational attainment and global publicity about polar environments and their wildlife provide both the means and the motives to travel to polar destinations. Competition among growing numbers of operators is progressively reducing costs of polar travel.
- *Improving transport and infrastructure.* Improvements in marine, air and ground transport technologies, e.g. ice-strengthened hulls, ice-breaking capabilities, improved navigational technologies, better hydrographic, ice and weather information and advances in telecommunications, all contribute to growth. The deployment from 1990 of Soviet-era icebreakers as cruise ships extended the tourist experience to the limits of both polar seas. Commercial air transport is increasing the frequency of its schedules, the fuel efficiency of its jets and the expansion of its itineraries to new Arctic destinations and southern hemisphere gateway cities for Antarctica. Land transport now routinely includes motor coaches and expanded rail services to transport growing numbers of tourists.

- *Growing peripheral support.* Arctic governments, and southern hemisphere countries with gateway ports offering services for Antarctic tourists, are providing new transport and support service infrastructure. Motivated by the economic benefits derived from tourism, significant investments in marine and airport expansions, improved ground transport, telecommunications and logistic support are either in place, under construction or being planned. Partnership arrangements between the cruise industry and polar governments, vigorously reinforced by both government and industry-sponsored promotional campaigns, can only help to stabilise and perpetuate growth.

As tourism in any area expands and diversifies, the effectiveness of its regulation is diluted, but yet becomes more critical. What worked for a few hundred visitors per year may prove ineffective when tens of thousands turn up to enjoy the same experiences. In the Arctic, regulating bodies receive constant feedback from operators; they are likely to know if numbers are getting out of hand in any particular sector, and may vary their regulations accordingly. There is no such mechanism for the Antarctic – no indigenous communities, no rangers or designated monitors – to report on effects of crowding, and no means currently in operation to effect remedies.

Growth beyond capacity?

Despite the magnitude of infrastructure investments, growth of tourism, especially of shipborne tourism, consistently outpaces infrastructure development – a matter of serious management concern. In the Arctic, infrastructure expansion contributes to growth, but cannot keep pace with navigational, emergency and support service needs. Governments' efforts to gain economic benefits from expansion have yet to be matched by the financial, administrative, infrastructure and personnel commitments required for effective management. This places both tourists and polar resources at risk, representing a potentially serious liability for governments and tour operators.

In the Antarctic, absence of governmental infrastructure has in the past been offset by semi-official mutual aid agreements between tour operators working through the International Association of Antarctica Tour Operators (IAATO) and national scientific stations – a system that worked satisfactorily when numbers of ships and voyages remained small. However, increasing numbers, plus the arrival on the scene of large cruise ships, makes these arrangements no longer workable. Even the largest scientific stations would find difficulty in helping a distressed cruise liner with 3000 or more passengers and crew.

Further strengthening of IAATO's emergency response protocols and government-sponsored search and rescue capacity are under way (DIRECTEMAR, 2008). Based on close cooperation between IAATO and the Search and Rescue agencies of several governments, serious attempts are being made to provide adequate responses to emergencies. However, there are no sustained sources of funding for emergency infrastructure facilities or other support services for Antarctica. To date, the lack of infrastructure has not been a barrier to Antarctic tourism because of the self-reliance of the tour operators. But the increasing frequency of serious marine incidents, including sinkings, cannot fail to intensify concerns about the continuing absence of infrastructure.

Warmer welcome

Legal barriers prohibiting entry to parts of the Arctic were until recently as formidable as the naturally occurring challenges of ice and cold. The Soviet Union exercised a right of sovereignty that effectively blocked access to the Arctic's largest land mass and the longest coastline of the Arctic Ocean. Military and national defence installations established by most other Arctic nations, either individually or jointly, prohibited public access to many locations.

The end of the Cold War heralded a new era for Arctic tourism. The Russian Federation's Ministry of Tourism now strongly encourages access to Arctic tourist venues of all kinds from the Kola Peninsula to Kamchatka. Cruise ships operate in the Kara, White, Barents and Bering seas, and transit the Northern Sea Route. Sport fishing, trophy hunting, heritage tourism and ecotourism are now promoted by the Russian Federation. Ironically, Cold War military installations that were once off limits have become tourist attractions. Infrastructure and facilities at several of those sites are vital components for the delivery of commercial tourism services.

Jurisdictional changes associated with native people's attainment of sovereignty have also contributed to the growth of Arctic tourism. The achievement of self-governance by native people focused serious attention on their need for economic self-sufficiency, toward which tourism in various forms is an important development strategy. Alaska Native Corporations, Inuit Home Rule Government in Greenland, the Nunavut in Canada, the Saami throughout their territory and more than 40 Russian indigenous peoples are actively promoting tourism within their domains. Federally established national protected areas dedicated to a variety of allowable recreational uses are also contributing to growth; all the Arctic nations are actively creating national parks, wildlife refuges, marine sanctuaries and wilderness areas that are major tourist attractions.

Geographic expansion of Arctic tourism has correspondingly increased the need for improved management. Motivated by prospects of economic benefits, governments need to be diligent in establishing regulations and management regimes to protect environmental and cultural resources, both directly and indirectly. Direct controls, for example, include officially designated land protected areas and marine sanctuaries that specify allowable resource uses, including recreation. Designation alone does not automatically safeguard those resources: their actual protection requires day-to-day management, including vigilant monitoring and enforcement.

Similarly, Arctic governments fully control the recreational uses of their wildlife and fisheries. Sport hunting and angling regulations can be readily applied to both land and marine resources within their jurisdictional boundaries. But again, actual conservation of those resources requires active management. Most recently, Arctic jurisdictions governed by indigenous people have facilitated a variety of heritage tourism developments. To sustain cultural integrity, governance must extend an equivalent level of effort for management.

Encouragement of cruise ship tourism by all Arctic governments creates its own set of governance and management challenges. Economic benefits arise, but serious regulatory and management problems follow. Many of the vessels are foreign-flagged and operating predominantly in international waters, making it difficult for Arctic jurisdictions to control operations and protect marine resources. The most readily available remedies include rigorous support for international treaties, polar codes that promote the use of vessels suitable for ice conditions, and vital emergency response practices that encourage safe operations. Ratification of international treaties, and improved definition of maritime jurisdictional boundaries, strengthen existing regulations and improve clarity. However, they must be underpinned by sound management objectives, and effected by a fleet of coast guard vessels capable of enforcing the regulations.

In summary, tourism's expanding geographic setting on land and sea, though encouraged by governments, intensifies problems of regulation and management that have yet to be matched with financial, administrative, infrastructure and personnel commitments. This places both tourists and polar resources at risk, and represents a potentially serious liability for governments and tour operators.

Multiple Resource Management Planning

These complexities of tourism regulation and management require a comprehensive approach that equally considers conservation of environmental and cultural resources, economic feasibility, social well-being of host communities and safety of tourists. Multiple resource management planning (MRMP) is a technique of proven success throughout the world, providing resource management planning that may be modified with time in response to changing conditions. Its principles and approaches can be applied on any scale from national parks to islands, beaches or communities. In polar tourism, its use has already been suggested in planning for tourism on South Georgia (Snyder, 2003; Snyder & Stonehouse, 2007) and to several situations requiring management planning in Antarctica (Bertram & Stonehouse, 2007).

We cite MRMP here, not as a catch-all system for solving polar tourism's problems, but as a systematic approach that addresses consideration of all the elements required in planning in both polar regions.

Figure 9.2 shows a model, based on Snyder's suggested management planning for South Georgia, which would be readily adaptable to larger or smaller areas at either end of Earth, with or without human populations.

The model requires recognition of five phases:

(1) Identifying critical elements.
(2) Defining the tourist experience.
(3) Project planning.
(4) Applying sustainability criteria.
(5) Implementation and operation.

Here, we consider its possibilities as a general approach to planning and management of polar tourism north and south, using Snyder's South Georgia study as an illustration and point of reference for the terms used.

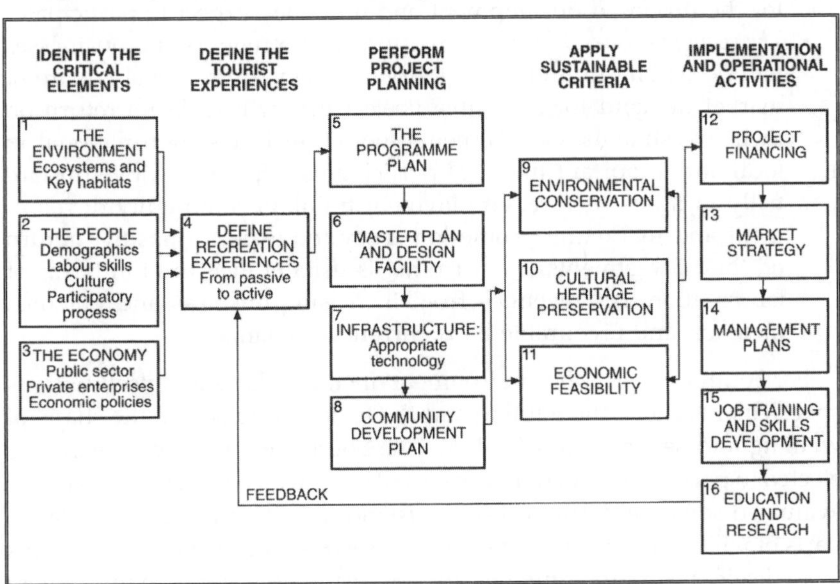

Figure 9.2 A systematic approach to tourism management, of general application to polar tourism. Each column represents a phase.
Source: Snyder (2003)

Identifying critical elements

Phase 1 of the MRMP considers the elements implicit in almost any plan involving tourism to a polar destination. Three critical elements identified in Boxes 1–3 of Figure 9.2, need to be taken into account:

- *The environment (Box 1).* This can be an ecological system of any size, forming a planning unit with recognisable environmental integrity, which is a tourist attraction and for which sustainability is required. Examples are a watershed, an island, an Antarctic landing site, a stretch of river or any other distinctive wildlife habitat. For MRMP purposes, the environment must be definable in terms of its distinguishing natural features and dynamic characteristics, and possess or include features deemed to be at risk from tourist activities that need to be brought under control.
- *The people (Box 2).* These are residents within or near the tourist attraction, who must be included in the planning and development of a tourism programme. Their support is essential for planning and implementing a successful tourism project. Both quantitative and qualitative data must be gathered on their views. Quantitative data include demographic and economic information, such as population size and age, race, ethnicity and occupational information. Qualitative information includes their cultural values, overall goals for the proposed development and how they expect to participate.
- *The economy (Box 3).* Private investors, entrepreneurs and businesses will be willing to undertake tourism if they are reasonably sure of market demand for a feasible project that will realise a return on their investments. Governments too, at all levels from national to local, are important financial participants, often investing substantially in infrastructure, employment training, community development and marketing promotion. They too expect a positive return on their investments, a return that is usually measured in terms of job creation, tax revenues from increasing business and personal incomes, and favourable hard currency exchange.

For South Georgia, 'the environment' includes the island itself with its spectacular scenery, the wildlife and the historic artefacts in the shape of whaling and sealing relics (p. 134). The South Georgia government has elected to permit tourism only at selected sites, which therefore need to be evaluated separately. The rest of the island may be given reserve status, parts of which may be opened later to tourism should the need arise. This is in itself a strong argument for localising tourism initially, where possible at sites that have already proved both robust and attractive.

In the Arctic, 'the environment' may be a designated landing site, or a settlement, village community, stretch of river or expanse of tundra set aside for hunting. In the Antarctic too, it could be a designated landing site, though no such sites currently exist. Passengers may be landed anywhere except in the precincts of scientific stations (unless permission has been granted) or at sites that are set aside for their scientific interest or significance. Though some 300 sites are listed as having been used for landings over the years. (Haase, personal communications), tour operators, in fact, land passengers at a relatively small number of sites, varying from year to year, which are notable for their wildlife, scenic or historic interest. Some of these have recently been inventoried, and guidelines have been issued for their protection. In MRMP terms, each could become a designated unit of 'the environment'.

On South Georgia, there are few people to take into account. There is no indigenous population – only a small community of government scientists and administrators, including the managers and caretakers of the small whaling museum. Much the same is true of the Antarctic region as a whole: scientists and support staff at research stations are the only 'people' likely to be met. However, their stake in the region needs to be respected, and their interests protected from intrusion. Throughout the Arctic, 'the people' are the many local communities, whose interests are paramount in any tourism venture, and best served by allocating to the people full status in decision making and planning.

'The economy' is the sum of costs and gains involved in tourism activities. On South Georgia, the government provides regulation and services, for use of which tour operators bring visitors and make payments. In the Arctic, similar arrangements exist between national and local governments, and a range of clients, from tour operators to individuals, who pay for services received and contribute through purchases, taxation and in other ways to national, local and private exchequers. Moneys so received are used to support and improve infrastructure, including rangers, safety and emergency measures, information, education, research, monitoring and all other services that help to improve the tourism experience. Profits make welcome contributions to standards of living in local communities.

Only in the Antarctic is 'the economy' lacking. The Antarctic Treaty System (ATS) has made no provision for taxing tourism. Tour operators use the whole area south of 60°S free of charge, with no benefits accruing either to the ATS itself, or to the region. The ATS has, in consequence, no income to provide services, to conduct research into tourism or its effects, or even to institute the most basic enquiries into the success of its own tourism-regulating policies.

Unfortunately, this curious situation is one that neither the ATS nor the tour operators, as represented by IAATO, seem at present concerned to change. This is understandable, because it would commit the ATS to developing a management organisation for tourism (perhaps one parallel to CCAMLR: p. 152), which has so far been deemed unnecessary and would probably require years of further discussion to establish. It would further commit the tour operators to increasing fares, which they too might deem unnecessary after a tax-free half-century.

Defining the tourist experience

The decisions made in Phase 1 of the MRMP determine in general terms what tourism experiences can be offered, extending perhaps from passive viewing of animals or scenery by land, sea or air modes of transport, to more participatory activities – hiking, backpacking, river rafting, kayaking, sport fishing or mountaineering. The surveys will indicate resident's abilities, capacities and willingness to lead, support and generally enrich these activities, and economics will identify ways in which they can be safely and efficiently provided.

With Phase 1 completed, Phase 2 requires regulating authorities and tour operators to examine further what forms of tourism are appropriate for permitting in the designated environments.

In relation to South Georgia, Snyder and Stonehouse comment:

> Arising from the fact that tourist visits began before the government saw reason to intervene, tour operators were for several years allowed to do more or less as they pleased, bound only by their own concepts (fortunately very appropriate ones) of environmentally sound behaviour. More active government intervention alters the rules: it is for operators to seek, and for managers [i.e. government] to provide, the opportunities and levels of permitting, and to determine the condition within which the activities take place. (Snyder & Stonehouse, 2007: 256)

Within Box 4, then, would be found a menu of permissible activities and the terms under which they may take place, agreed mutually between regulators and operators. Such agreement must be deemed an essential part of any contract between those with duties to protect resources and those who seek to use them. For the Arctic, any tour operator seeking to introduce clients to a new venue should have no difficulty in identifying the appropriate authority. There may be difficulties or disagreements in negotiating usage, particularly where more than one government department or authority holds a stake in management, but at least no ambiguity over responsibilities once agreement has been reached. For the Antarctic, operators are required to seek clearance from their own governments, which base decisions ultimately on the not-always-clear wording of the Protocol on Environmental Protection to the Antarctic Treaty.

Project planning

The third phase of the MRMP is to create written plans, which satisfy both regulators and operators, that proposed projects are viable and organised to their mutual satisfaction. The model provides for four kinds of plan, any or all of which may be appropriate throughout the polar regions. These tasks are briefly described below and appear in Boxes 5–8 of Figure 9.2.

- *Tourism programme plan (Box 5*). This kind of plan provides definitive statements regarding the ways in which the tourism operations will be conducted. It incorporates ways in which tourism experiences will be delivered, including modes of transport, specialised equipment, support personnel, emergency response services, facilities, infrastructure and other resources needed to conduct operations. Information generated by these tourism plans identifies essential resources and estimated costs for delivering guest services, as well as providing valuable inputs to other planning tasks.
- *Master plan (Box 6)*. This determines the most appropriate land, water and cultural resource uses for proposed tourism regions and sites. Its actual size, design, support systems and amenities will be derived from ways in which a foregoing tourism programme plan defines the array of experiences to be offered. Additional considerations are that facility and support service designs should be consistent in size and character with the environmental and cultural setting. Estimated costs of facility construction, operations and maintenance are included within the master plan.
- *Infrastructure plan (Box 7*). Produced cooperatively between tourism operators and the governing jurisdiction where tourist activities will occur, this is essential because polar tourism projects are frequently located in remote or very rural locations, with minimal infrastructure and difficult conditions. Sizing and selection of infrastructure must be compared with costs of providing services. Tour operators cannot afford the financial burden of a major utility system, but nonetheless are responsible for providing sanitary and other essential services to their guests. The government provides infrastructure to encourage tourism and to meet those needs, but its finances are usually severely constrained. Ultimately, tourism needs to demonstrate environmental responsibility, and the government needs to support the community and economic development it is encouraging by improving services for its citizens. Reconciling those difficult trade-offs is best resolved at the outset, through a cooperatively determined infrastructure plan.
- *Community development plan (Box 8*). Arising from involvement of local populations affected by a proposed tourism development

project, this must show how the project is likely to enhance the social and economic well-being of the community. When thoughtfully planned, tourism projects can serve as a positive catalyst for improving the quality of life in a community.

Current operations on South Georgia and Antarctica, mainly involving brief landings from ships to visit wildlife and historic sites, require few or no government resources. Snyder and Stonehouse (2007: 259) deem some of the above (e.g. a community development plan) unnecessary. However, visits to Arctic reserves and settlements requiring more use of infrastructure – employing, for example, guides, rangers and facilities – stand more in need of such plans.

On South Georgia, where there has been much recent research on heritage resources (in particular the old whaling stations), the master plan includes notification of which of these can be included in tourist activities, and how they may be used. For the Arctic, management agencies responsible for sites will require such plans to attract tourism and facilitate smooth running. There is no equivalent agency to prepare such plans for Antarctic visited sites.

Community development plans (Box 8) are of particular importance where indigenous populations are involved, to ensure that any proposed tourism runs in accord with long-term community development objectives. Specifically, they should be the positive result of the involvement of the local populations in the development project. They are valuable because they propose realistic techniques for enhancing the social and economic well-being of the community, and are catalysts in improving quality of life.

Sustainability criteria

Phase 4 of the MRMP provides for conservation. Today, planning for tourism in any environment should be based on two key propositions – that sustainability is an essential component (p. 85), and that sustainability factors built into a plan must be testable. Boxes 9–11 provide for environmental conservation, cultural heritage preservation and economic feasibility. The first two take care of important resource protection: the third ensures that both are adequately funded. Each is briefly described as follows:

- *Environmental conservation (Box 9).* This lays down rules for the sustainable use of wildlife, scenic beauty and other factors that have been catalogued earlier in the planning process. The criteria require an understanding of the tolerances of the flora and fauna to contact with humans, derived from an inventory of biological species involved, knowledge of the environmental dynamics, and of the natural hazards to which the sites and their inhabitants are subjected. From these factors can be determined the overall tolerance of sites to human usage.

- *Cultural heritage preservation (Box 10).* Herein are listed the special needs of communities and heritage sites at risk. Host communities must be able to continue their customs, religion and social practices authentically, and criteria for sustainability must be derived from the local population. How many visitors can be accommodated at a time? How can local people conduct tours and participate economically in other ways? Can education be provided to maintain indigenous language and cultural practices? What additional expansion or improvement of local infrastructure is needed to satisfy both local communities and tourists?
- *Economic feasibility (Box 11).* This box recognises that all processes involved in securing sustainable tourism, from initial planning to final monitoring, are serious matters to be undertaken to professional standards, and likely to be costly. To be effective, administrators must also be alert to changes in tourist fashions and preferences, i.e. make provision for market analysis. As services provided by administration – whether at national or local levels – these are a legitimate charge on revenues from tour operators, and ultimately from the tourists themselves. From a strictly monetary point of view, economic feasibility is thus a matter of ensuring that revenues from tourism are adequate to provide services, and that the services are adequate to ensure that the tourism is sustainable – a point considered more fully in Phase 5. Other important aspects of economic feasibility include the jobs, household income and local business revenues derived from tourism. Collectively, the total benefits of tourism must exceed the costs.

To be effective, evaluation of Phase 4 criteria demands more than a cursory listing of possible hazards, and periodic inspections to 'monitor' if changes are occurring. For species, artefacts or systems at risk, a more rigorous approach is to predict what changes are likely to occur, set limits of tolerance to change as an integral part of the management plan, and control visitor pressures to remain within the imposed limits. 'Monitoring' then becomes more than the casual observation that satisfies legal requirements. It assumes its proper ecological rôle (pp. 95–97) of ensuring that stated management objectives are being achieved, or if not, that adjustments are needed to maintain required standards of integrity.

Implementation and operation

Phase 5 of the MRMP develops the theme of Box 11, ensuring that operations performed under the plan are financially as well as environmentally viable. Snyder and Stonehouse characterise this phase:

Operational activities are perpetually challenged to strike a balance between allowable human uses and the resiliency of environmental and cultural resources to accommodate those uses. All of this must be viewed by tour operators and local government as economically feasible, or there will be no tourism whatever. Phase 5 boxes 12–16 are reminders of business concerns that are likely to need attention from time to time. (Snyder & Stonehouse, 2007: 259–260)

The boxes are labelled and defined as follows:

- *Project financing (Box 12)* requires (1) a knowledge of the anticipated costs and revenues associated with a tourism project, (2) the identification and pre-qualification of the sources of financing, (3) a plan for acquiring both the equity and debt financing to implement the project. All Arctic governments play a critical rôle in providing sources of tourism project financing, often combining direct funding, infrastructure financing, tax credits, personnel training and other forms of financial support. Governmental participation has proven crucial for accomplishing much of the economic development Arctic communities seek, and the support that tour operators need to conduct feasible businesses.
- *Marketing programmes (Box 13)* take account of the tourism markets identified in the tourism programme plan. In several instances, Arctic governments produce tourism development plans for the purpose of identifying specific target markets. Marketing programmes in the Arctic are highly coordinated between governments and tour operators. Travel schedules for cruise ship port arrivals and departures, commercial air operations, railroads and coastal ferries are thoroughly coordinated between governments and operators, in order to maximise economic opportunities and minimise infrastructure capacity constraints. For those seeking more effective management, careful evaluation of the highly integrated marketing programmes would be beneficial. Such an examination would, for example, quickly reveal the number and type of cruise vessels, the number of passengers, times of arrival and the capacities of the host community.
- *Management plans (Box 14)* are produced to establish guidelines for the efficient and effective performance of visitor services, facility operations, community outreach and environmental stewardship. These may take several forms to include, for example, environmental management, wildlife, marine, cultural resources, recreation resources and hospitality service plans.
- *Job training and skills development (Box 15)* are important for the sustainability of the tourism project and the host community. Tourism operations benefit from well-trained personnel who

provide quality services to the guests, with low rates of turnover and absenteeism. Education and skills development for the local people should contribute to their realisation of human potential and enable them to become economic beneficiaries of the project. Training and skills development are essential to any growing enterprise: the Arctic in recent decades has met the challenge of developing a workforce, from rangers to waiters, coach drivers and bush pilots, to meet the needs of thousands of tourists with money to spend and leisure to enjoy.

- *Education and research (Box 16)* provide benefits to tour operators by expanding their market to include educational institutions, enriching the tourism experience of their guests and identifying changes in environmental conditions that may affect their operations. Based on competent environmental research, tourism operations can be monitored and evaluated by affected government agencies. This information serves as a feedback mechanism for regularly evaluating the quality and integrity of both environmental quality and the tourism experiences. Education and research are needed as reminders that, for at least a proportion of visitors, 'leisure' implies learning; the discovery and imparting of information is perhaps the least harmful motivation for promoting and encouraging tourism in wilderness areas.

Though for any particular project, one or several of these headings may prove redundant, together they provide a framework for creating practical operational standards, and harmonious relations between administrators, operators, tourists and involved communities. Project financing and market strategy (Boxes 12 and 13) provide for details of projected costs and revenues from particular tourist operations, and anticipating what opportunities a growing market may provide for further business. Management plans (Box 14) are not static: they may need frequent revision to keep up with market trends and developments.

Phase 5 of this MRMP outline is a reminder that tourism is a cooperative venture – a form of partnership between regulators and entrepreneurs. The cooperation may be willing, as it now appears to be over much of the Arctic, and indeed on South Georgia. Both partners are keen to see further development within mutually recognised boundaries. Where cooperation is sound, tourism advances in ways and at a pace agreed by both, to the benefit of both.

In the Antarctic, by contrast, the ATS regulators are reluctant partners with entrepreneurs, unwilling or unable to accept the equivalent responsibilities of partnership, and gaining from it no financial benefits that might encourage them – and would certainly allow them – to play their rôle more fully. Keenness is limited to the entrepreneurs, who alone profit from the enterprise. Tourism still advances, because there is nothing beyond market demand to constrain it. The pace of advance is determined, not by mutual agreement between partners, but only by market forces that the ATS has no powers to control, and the industry gains no advantage from controlling.

If the kinds of planning exemplified in the MRMP illustrated above seem excessive or unnecessary to the ATS regulators, and beyond their means of achievement, they may be sure that such planning is part of the stock-in-trade of their more business-minded partners, and could help to explain the industry's current ascendency.

Summary and Conclusions

Increasing numbers of visitors and increasing diversity of services, over a century of tourism in the Arctic and half a century in the Antarctic, demonstrate beyond doubt the lasting attractions of those regions. The natural obstacles that formerly inhibited polar tourism have been surmounted; there is no shortage of clients with time and means to travel, and the industry is set for further expansion at both ends of Earth.

Growing popularity is accompanied by growing perils. More cruise ships, voyages and flights each year, more marine incidents endangering cruise ships, growing numbers of adventurers pursuing high-risk recreation in polar regions: these are some of the more dangerous components of growth. Other negative impacts, all too common with ill-controlled tourism elsewhere in the world, are starting to appear. Overcrowding at attractive sites, wear-and-tear appearing in cherished wilderness, indigenous communities divided by ill-designed heritage tourism, unable to determine what constitutes 'appropriate' numbers of visitors: these too are dangerous in destroying the priceless foundation resources – the ultimate reasons for the existence of polar tourism.

Growing perils and negative impacts can be minimised by sound regulation and management, both of which are needed north and south, to safeguard the environmental and cultural integrity of polar regions, and to safeguard the many thousands of tourists who now visit them annually.

Arctic governments, while eagerly seeking the economic benefits of tourism, acknowledge their management and infrastructure deficiencies, and are now pursuing a 'catch-up' approach to tourism regulation and management. Antarctica's system of government, with no incentives to welcome tourism or promote its expansion, seeks only to contain it by regulations that apply to all human activities in the area. Many tour operators at either end of the world are sensitive to negative impacts and seek to mitigate them, but have neither powers nor ability to calculate, much less manage, the cumulative impacts that must arise from long-term usage of the polar regions' resources.

The case for more effective regulation and management both north and south is self-evident. We offer MRMP as a tool for better understanding, evaluating and managing the problems of sustainable polar tourism.

The existence of polar tourism at both ends of the world is a triumph of human will over enormous environmental, economic and technological obstacles. Managing its growth and diversity will require no less an amount of determination and commitment.

Appendix A

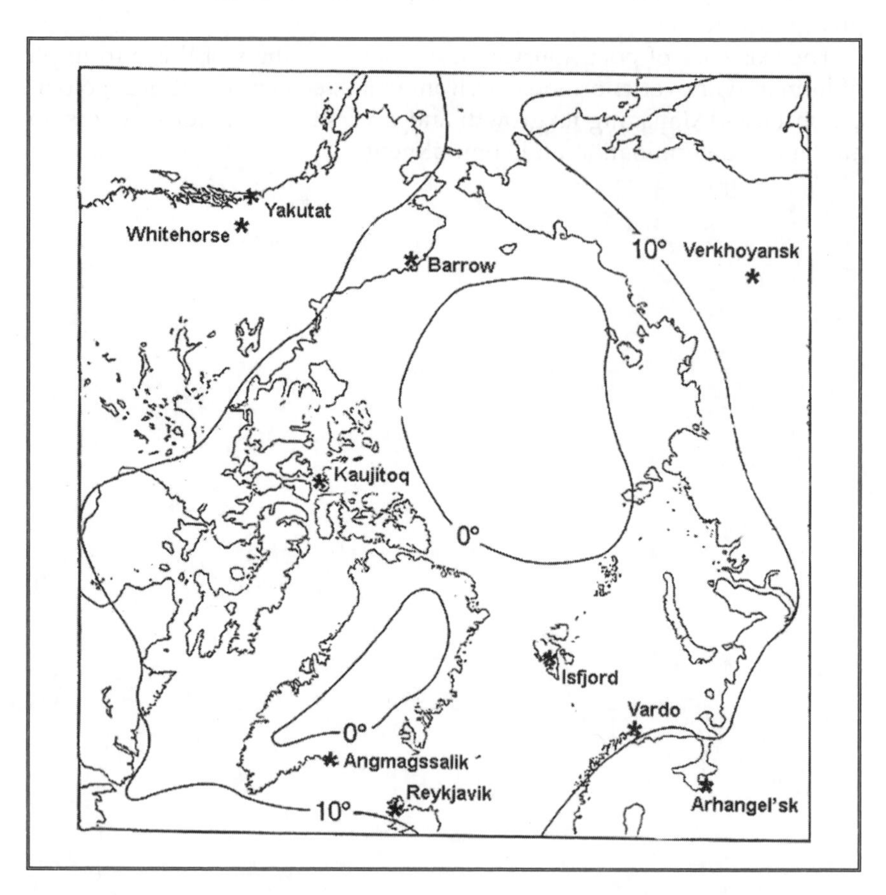

Figure A.1 Arctic stations

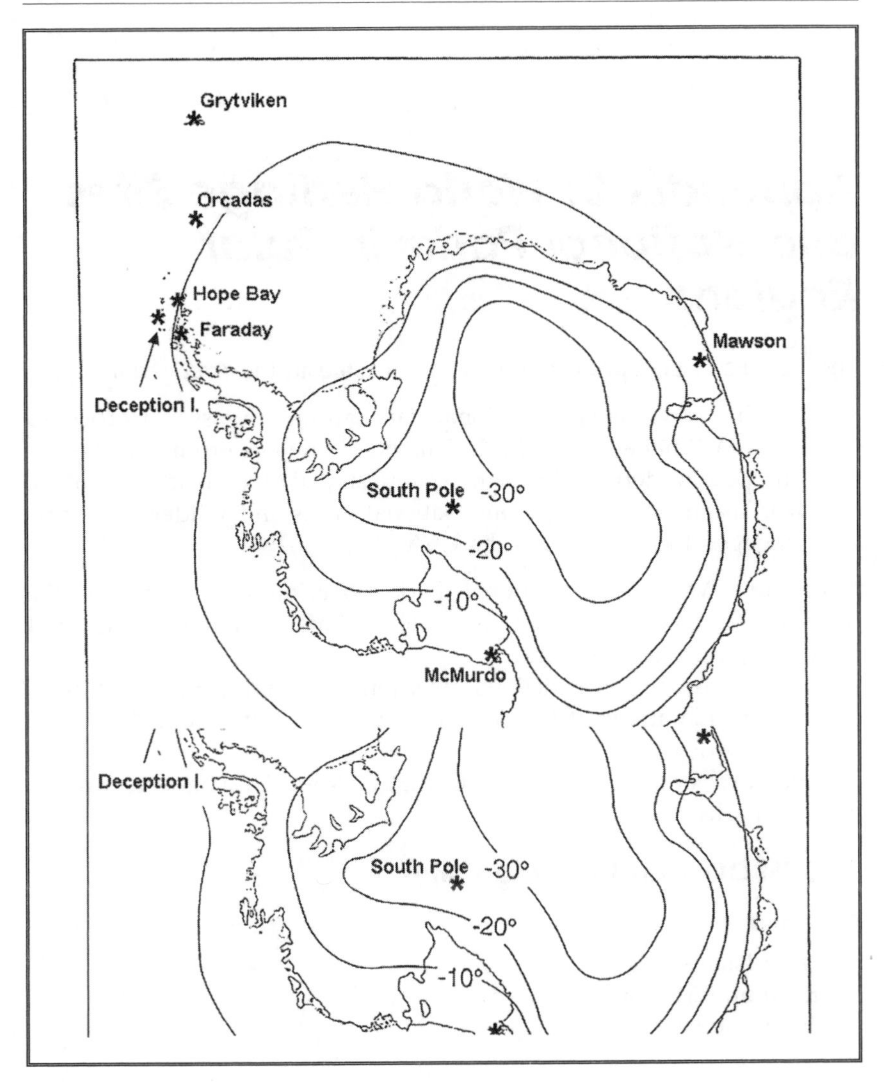

Figure A.2 Antarctic stations

Appendix B: World Heritage Sites and National Parks in Polar Regions

The world's largest protected areas are located in the Arctic, notably:

- North-East Greenland National Park, an enormous area of icecap and coast managed by the Greenland Home Government;
- Europe's Wilderness Area, an expanse shared by Finland and Russia;
- North America's contiguous National Parks and Wilderness Areas managed by Canada and the USA;

Iceland's Vatnajökull National Park, covering one eighth of the country's surface and including the Skaftafell National Park and Jökulsárgljúfur National Park.

All of the Arctic's jurisdictional decisions to establish protected areas involve considerations of the economic development potential of those areas for tourism.

Antarctica south of 60°S has no UNESCO World Heritage Sites and no national parks.

A. UNESCO World Heritage Sites

Australia

Fraser Island
Heard and McDonald Islands
Macquarie Island

Canada

L'Anse Aux Meadows National Historic Site
Nahanni National Park Preserve
Dinosaur Provincial Park
Sgaang Gwaii (Knight Island)
Head Smashed In Buffalo Jump
Wood Buffalo National Park
Canadian Rocky Mountain Parks
Historic District of Quebec
Gros Morne National Park

Old Town Luenburg
Miguasha Park

Canada and the USA

Kluane/Wrangell – St Elias/Glacier Bay/Tatshenshini – Alsek
Waterton – Glacier International Peace Park

Finland

Old Rauma
Fortress of Suomenlina
Petajavesi Old Church
Verla Groundwood and Board Mill
Bronze Age Burial Site of Sammallhdenmaki

New Zealand

New Zealand Sub Antarctic Islands

Norway

Urnes Stave Church
Bryggen
Roros
Rock Drawings of Alta

Russian Federation

Cultural and Historic Ensemble of the Solovetskiye Island
White Monuments of Vladimir and Suzdal
Volcanoes of Kamchatka

Sweden

Royal Domain of Drottningholm
Birka and Hovrgarden
Englesberg Ironworks
Rock Carvings in Tanum
Skogskyrkogarden
Hanseatic Town of Visby
Church Village of Gammelstad, Lulea
Laponian Area
Naval Port of Karlskrona
High Coast
Agricultural Landscape of Southern Oland
Mining Area of the Great Copper Mountain in Falun

B. National Parks

Finland

Aulanka National Park
Ekenas Archipelago National Park
Helvetnjarvi National Park
Koli National Park
Kolovesi National Park
Lemmenjoki National Park
Linnansaari National Park
Oulanka National Park
Palas-Ounastunturi National Park
Patvinsuo National Park
Petkeljarvi National Park
Puurijarvi-Isosuo National Park
Pyhatunturi National Park
Urho Kekkonen National Park

Greenland

Northeast Greenland National Park
Lauge Kyst Nature Reserve

Iceland

Jokulsargljufur National Park
Skaftafell National Park
Pingvellir National Park

Norway

Borgefjell National Park
Dovrefjell National Park
Femundsmarka National Park
Forlandet National Park
Gressamoen National Park
Gutilia National Park
Hardangervidda National Park
Jostelsbreen National Park
Jotunheimen National Park
Nordvest Spitsbergen National Park
Ormtjernkampen National Park
Rago National Park
Reisa National Park
Rondane National Park
Altfjellet-Svartisen National Park

Stabbuedalen National Park
Sor Spitsbergen National Park
Ovre Anarjakka National Park
Ovre Divdal National Park
Ovre Pasvik National Park
Anderdalen National Park

Sweden

Abisko National Park
Stora Sjofallets National Park
Sarek National Park
Padjelanta National Park
Muddus National Park
Peljekaise National Park
Skula Skogen National Park
Goska Skandon National Park

USA

Aniachak National Monument and Preserve
Bering Land Bridge National Preserve
Cape Krusenstern National Monument
Denali National Park
Gates of the Arctic National Park
Glacier Bay National Park and Preserve
Katmai National Park
Kenai Fjords National Park
Klondike Gold Rush National Historic Park Alaska
Kobuk Valley National Park
Lake Clark National Park
Noatak National Park
Sitka National Historic Park
Wrangell St Elias National Park and Preserve
Yukon-Charley Rivers National Preserve

Appendix C: Maritime Conventions

In relation to Chapter 6, this listing summarises the fields of responsibility of 10 conventions concerned with maritime operations, in chronological order.

The 1969 Convention Relating to Intervention on the High Seas in Cases of Oil Pollution Casualties

This convention, which was adopted by the International Maritime Organisation (IMO) in 1969 and entered into force in 1975, affirms the right of coastal states to take action on the high seas to protect their coastline or related interests from pollution by oil, following a collision or other maritime casualty. A Protocol added in 1973, and later amendments in 1991, 1996 and 2002, extended the convention to cover chemicals and other noxious substances. Before acting, the coastal state should notify the flag state of the ship, consult independent experts and notify any person whose interests may reasonably be expected to be affected by their action, though in cases of extreme urgency, measure might be taken at once.

The 1970 Arctic Waters Pollution Prevention Act

The Arctic Waters Pollution Prevention Act was enacted by the Canadian Government to prevent the pollution of waters next to the mainland and islands of the Canadian Arctic by boats and ships. Regulations arising from the Act govern the deposit of domestic and industrial waste, the compulsory reporting of a deposit of waste that violates the regulation, certificates of insurance that must be provided by ship owners/operators and limits of liability for ships with insurance. They also govern construction standards for ships passing through Arctic waters, bunkering stations, certificates that owners or operators must obtain before taking their ship through certain Arctic waters, the use and qualifications of ice navigators (i.e. persons authorised to guide ships through certain areas), supplies of fuel and water in Arctic waters and authorisations and standards for sewage and oil deposits from ships.

The 1972 Convention on Prevention of Marine Pollution by Dumping of Wastes and Other Matter (London Convention)

The London Convention sought to prevent pollution of the sea by permitting dumping of wastes, except for those on a banned list. The convention was strengthened in November 1996 by the addition of a Protocol that prohibited dumping, except for materials on an approved list. Permitted materials include dredgings, sewage sludge, fish wastes, vessels, platforms or other man-made structures, inert, inorganic geological materials, organic materials of natural origin and certain bulky items of inert materials that isolated communities cannot reasonably dispose of elsewhere.

The 1973 International Convention for the Prevention of Pollution from Ships as modified by the 1978 Protocol (MARPOL 73/78)

The International Convention for the Prevention of Pollution from Ships was adopted on 1973, covering pollution by oil, chemicals, harmful substances in packaged form, sewage and garbage. In 1978, at a Conference on Tanker Safety and Pollution Prevention, a new Protocol was adopted, absorbing the parent convention. The combined instrument, known as MARPOL 73/78, includes regulations aimed at preventing and minimising pollution from ships – both accidental pollution and that from routine operations. It currently includes six technical Annexes:

- the prevention of pollution by oil;
- the control of pollution by noxious liquid substances in bulk;
- the prevention of pollution by harmful substances in packaged forms;
- the prevention of pollution by sewage from ships;
- the prevention of pollution by garbage from ships;
- the prevention of air pollution by ships.

Annex 1 specifies the total quantity of oil that a tanker may discharge in any ballast voyage while under way, the rate at which the oil may be discharged and that none whatsoever may be discharged from cargo spaces within 50 miles of the nearest land. Recent amendments require new tankers to have double hulls, and ballast tanks that do not require discharge of oil. The remaining annexes prescribe measures similarly designed to minimise pollution from discharges, including special sea areas (e.g. the Baltic Sea) in which no discharges are allowed.

The 1974 International Convention for the Protection of Life at Sea (SOLAS)

Though the current version is dated 1974, SOLAS originated in 1914 as an international response to the loss of *Titanic*. It was revised in 1929, 1948 and 1960. The new convention adopted in 1974 included all the amendments agreed up to that date, and introduced a new and faster amendment procedure – one that allowed an amendment to enter into force on a specified date unless, before that date, objections had been received from an agreed number of Parties. The 1974 convention, updated and amended frequently, is generally referred to as 'SOLAS, 1974, as amended'.

The most recent amendments, adopted by the IMO in December 2006 and expected to come into effect by 2010, include provision of:

- safe areas and essential systems to be maintained while a ship proceeds to port after a casualty, which will require redundancy of propulsion and other essential systems;
- on-board safety centres, from where safety systems can be controlled, operated and monitored;
- fixed fire detection and alarm systems, including requirements for fire detectors and manually operated call points to be capable of being remotely and individually identified;
- fire prevention, including enhancing the fire safety of atriums, the means of escape in case of fire and ventilation systems;
- time for orderly evacuation and abandonment, including requirements for the essential systems that must remain operational in case any one main vertical zone is unserviceable due to fire.

Such amendments would represent considerable advances on the designs of some passenger cruise ships currently operating in polar waters, where any on-board emergency has heightened potential for becoming a disaster.

The 1982 United Nations Convention on the Law of the Sea (UNCLOS)

UNCLOS represents the United Nations' concern to govern all aspects of the world's oceans, including environmental matters, economic and commercial activities and settlement of disputes. It defines the sovereign rights of coastal states to natural resources and certain economic activities within a 200-nautical mile exclusive economic zone (EEZ), and to exercise jurisdiction for research and protection over the adjacent continental shelf (the national area of the seabed) for exploration and exploitation. The shelf can extend at least 200 nautical miles from the shore, and more in specified circumstances. While all states enjoy traditional freedoms of navigation, overflight, scientific research and fishing on the high seas, they are obliged to cooperate with other states in managing and conserving living resources. States bordering enclosed or semi-enclosed seas are expected to cooperate in managing living resources, environmental and research policies and activities. All states are bound to prevent and control marine pollution and are liable for damage caused by violation of their international obligations to combat such pollution. Part XII of the convention concerns protection and preservation of the marine environment, seeking to prevent, reduce and control pollution from land-based sources, from sea-bed activities subject to national jurisdiction or from such human activities as dumping and waste disposal. UNCLOS embodies a host of critical issues regarding the governance of Arctic seas, resolution of jurisdictional disputes, prevention of marine pollution and the protection of life at sea. However, the effectiveness of those measures will be severely restricted until all Arctic nations, most notably the USA, ratify the treaty.

The 1990 Oil Pollution Preparedness, Response and Cooperation Convention (OPRC)

This convention, in which the IMO plays an important coordinating rôle, requires Parties to establish measures for dealing with oil pollution incidents, either nationally or in cooperation with other countries. Their ships and offshore units are required to carry oil pollution emergency plans designed to provide a prompt and effective response to oil pollution incidents. Ships are also required to report incidents of pollution to coastal authorities, and the convention details actions that are then to be taken. The convention calls for the establishment of stockpiles of oil spill combating equipment, the holding of oil spill combating exercises and the development of detailed plans for dealing with pollution incidents. Parties are required to assist others in the event of a pollution emergency.

A Protocol on Preparedness, Response and Cooperation to Pollution Incidents by Hazardous and Noxious Substances, prepared in 2000 (OPRC-HNS Protocol) entered into force on 14 June 2007. Hazardous and noxious substances are substances other than oil that, if introduced into the marine environment, are likely to create hazards to human health, to harm living resources and marine life, to damage amenities or to interfere with other legitimate uses of the sea. The HNS Protocol will ensure that ships carrying hazardous and noxious liquid substances are covered by preparedness and response regimes similar to those already in existence for oil incidents.

The 1991 Arctic Environmental Protection Strategy (AEPS)

Ministers from the eight Arctic states (Canada, Denmark (Kalaallit Nunaat), Finland, Iceland, Norway, Sweden, the Russian Federation, and the USA) in June 1991 signed a declaration endorsing a joint strategy for the protection of the Arctic environment. The AEPS represents a collective approach to protecting, enhancing and restoring the Arctic environment and using natural resources sustainably. The strategy focuses on four issues:

- protection of the marine environment;
- emergency preparedness and response;
- conservation of flora and fauna;
- monitoring and assessment of contaminants.

In 1993, the ministers set up an Indigenous Peoples' Secretariat to facilitate communication and enhance the participation of indigenous peoples in the AEPS, and established a Task Force on Sustainable Development and Utilization. An Arctic Monitoring and Assessment Programme has been established to examine pollution and other environmental issues.

The 1995 Washington Declaration and the Global Programme of Action for the Protection of the Marine Environment from Land-Based Activities

This intergovernmental programme recognises threats to the health, productivity and biodiversity of the marine environment resulting from human activities on land. Municipal, industrial and agricultural wastes and run-off, as well as atmospheric deposition adversely affect estuaries and coastal waters. Human pressures on the coastal systems require action at all levels: local, national, regional and global. The United Nations Environmental Programme coordinates the Global Programme of Action, strengthening collaboration between agencies with mandates relevant to the impact of land-based activities on the marine environment.

The 2002 International Code of Safety for Ships in Polar Waters (Polar Code)

Developed from a Canadian initiative, and adopted by the IMO in 2002, these guidelines form the first internationally recognised standards for the construction and operation of ships in Arctic ice-covered polar waters. Often referred to as the 'Polar Code', they address vessel structure, maritime equipment, ship systems and operations, and take a comprehensive approach to issues of safe navigation and the prevention of pollution in polar waters, where lack of infrastructure and severe climatic conditions are key hazards. They point out, for example, that fire suppression systems should be designed so as not to be made inaccessible by accumulation of ice or snow, provide rules for personal and group survival kits and require all lifeboats carried by polar class ships to be fully enclosed: those not of polar class should carry tarpaulins to provide complete lifeboat covers. All ships operating in Arctic ice-covered waters should carry an operational manual and a training manual for all ice navigators on board, and ice navigator should have completed an approved training programme.

Adoption of the Code by the IMO represents progress, but as of June 2008 it remains a set of recommendations, which no state has made mandatory by legislation. No internationally approved training courses exist for ice navigators or masters operating vessels in ice-covered waters. For a critical review of the Code and its acceptance, see Jensen (2008).

Appendix D: Best-Practice Guidelines for Polar Visitors and Tour Operators

1. Antarctic Guidelines

The following statement formed part of an annex to Recommendation XVIII-1, adopted at the 18th Antarctic Treaty Consultative Meeting in Kyoto, 1994. *Guidance for visitors to the Antarctic* was accompanied by a longer document – *Guidance for those organising and conducting tourism and non-governmental activities in the Antarctic* – which is available on the Antarctic Treaty website.

Guidance for visitors to the Antarctic

Activities in the Antarctic are governed by the Antarctic Treaty of 1959, and associated agreements, referred to collectively as the Antarctic Treaty system. The Treaty established Antarctica as a zone of peace and science.

In 1991, the Antarctic Treaty Consultative Parties adopted the Protocol on Environmental Protection to the Antarctic Treaty, which designates the Antarctic as a natural reserve. The Protocol sets out environmental principles, procedures and obligations for the comprehensive protection of the Antarctic environment, and its dependent and associated ecosystems. The Consultative Parties have agreed that, pending its entry into force, as far as possible and in accordance with their legal system, the provisions of the Protocol should be applied as appropriate.

The Environmental Protocol applies to tourism and non-governmental activities, as well as governmental activities in the Antarctic Treaty Area. It is intended to ensure that these activities do not have adverse impacts on the Antarctic environment, or on its scientific and aesthetic values.

This **Guidance for Visitors to the Antarctic** is intended to ensure that all the visitors are aware of, and are therefore able to comply with, the Treaty and the Protocol. Visitors are, of course, bound by national laws and regulations applicable to activities in the Antarctic.

(A) Protect Antarctic wildlife

Taking or harmful interference with Antarctic wildlife is prohibited except in accordance with a permit issued by a national authority.

(1) Do not use aircraft, vessels, small boats, or other means of transport in ways that disturb wildlife, either at sea or on land.
(2) Do not feed, touch, or handle birds or seals, or approach or photograph them in ways that cause them to alter their behavior. Special care is needed when animals are breeding or molting.
(3) Do not damage plants, for example by walking, driving, or landing on extensive moss beds or lichen-covered scree slopes.
(4) Do not use guns or explosives. Keep noise to the minimum to avoid frightening wildlife.
(5) Do not bring non-native plants or animals into the Antarctic such as live poultry, pet dogs and cats or house plants.

(B) Respect protected areas

A variety of areas in the Antarctic have been afforded special protection because of their particular ecological, scientific, historic or other values. Entry into certain areas may be prohibited except in accordance with a permit issued by an appropriate national authority. Activities in and near designated Historic Sites and Monuments and certain other areas may be subject to special restrictions.

(1) Know the locations of areas that have been afforded special protection and any restrictions regarding entry and activities that can be carried out in and near them.
(2) Observe applicable restrictions.
(3) Do not damage, remove, or destroy Historic Sites or Monuments or any artifacts associated with them.

(C) Respect scientific research

Do not interfere with scientific research, facilities or equipment.

(1) Obtain permission before visiting Antarctic science and support facilities; reconfirm arrangements 24–72 hours before arrival; and comply with the rules regarding such visits.
(2) Do not interfere with, or remove, scientific equipment or marker posts, and do not disturb experimental study sites, field camps or supplies.

(D) Be safe

Be prepared for severe and changeable weather and ensure that your equipment and clothing meet Antarctic standards. Remember that the Antarctic environment is inhospitable, unpredictable, and potentially dangerous.

(1) Know your capabilities, the dangers posed by the Antarctic environment, and act accordingly. Plan activities with safety in mind at all times.
(2) Keep a safe distance from all wildlife, both on land and at sea.
(3) Take note of, and act on, the advice and instructions from your leaders; do not stray from your group.
(4) Do not walk onto glaciers or large snow fields without the proper equipment and experience; there is a real danger of falling into hidden crevasses.
(5) Do not expect a rescue service. Self-sufficiency is increased and risks reduced by sound planning, quality equipment, and trained personnel.
(6) Do not enter emergency refuges (except in emergencies). If you use equipment or food from a refuge, inform the nearest research station or national authority once the emergency is over.
(7) Respect any smoking restrictions, particularly around buildings, and take great care to safeguard against the danger of fire. This is a real hazard in the dry environment of Antarctica.

(E) Keep Antarctica pristine

Antarctica remains relatively pristine, the largest wilderness area on Earth. It has not yet been subjected to large scale human perturbations. Please keep it that way.

(1) Do not dispose of litter or garbage on land. Open burning is prohibited.
(2) Do not disturb or pollute lakes or streams. Any materials discarded at sea must be disposed of properly.
(3) Do not paint or engrave names or graffiti on rocks or buildings.
(4) Do not collect or take away biological or geological specimens or man-made artifacts as a souvenir, including rocks, bones, eggs, fossils, and parts or contents of buildings.
(5) Do not deface or vandalize buildings, whether occupied, abandoned, or unoccupied, or emergency refuges.

2. Arctic Guidelines

World Wildlife Fund Arctic: Principles and Codes of Conduct

The WWF International Arctic Programme sees tourism as one way to support the protection of the Arctic environment. According to WWF Arctic, tourism can be conducted responsibly so that visitors learn to appreciate and respect Arctic nature and cultures, as well as provide additional income to local communities and traditional lifestyles (www.panda.org, 2007).

Recognising both the positive and negative potential of tourism, in 1995 WWF Arctic began developing principles and codes of conduct for Arctic tourism, and a mechanism for implementing them. The goal was to encourage the development of a type of tourism that protected the environment as much as possible, educate tourists about the Arctic's environment and peoples, respect the rights and cultures of Arctic residents, and increase the share of tourism revenues that go to northern communities. WWF Arctic believes that the development of this type of tourism is in the best interest not only of conservation, but also of residents, business and government.

The Principles and Codes for Arctic Tourism were developed in cooperation between WWF Arctic, tour operators, conservation organisations, managers, researchers and representatives from indigenous communities during workshops held on Svalbard in 1996 and 1997. Participants developed a List of Potential Benefits and Potential Problems of Arctic Tourism, Ten Principles for Arctic Tourism, a Code of Conduct for Tour Operators, and a Code of Conduct for Tourists.

The Ten Principles for Arctic Tourism

(1) Make tourism and conservation compatible
(2) Support the preservation of wilderness and biodiversity
(3) Use natural resources in a sustainable way
(4) Minimise consumption, waste and pollution
(5) Respect local cultures
(6) Respect historic and scientific sites
(7) Arctic communities should benefit from tourism
(8) Trained staff are the key to responsible tourism
(9) Make your trip an opportunity to learn about the Arctic
(10) Follow safety rules

The *Code of Conduct for Tour Operators in the Arctic* and the *Code of Conduct for Arctic Tourists* are identical with the Ten Principles, except at Point 8, which invites operators to educate staff, and tourists to choose tours with well-trained, professional staff.

Natural Habitat Adventures: World Wildlife Fund Conservation Travel Provider

The effectiveness of WWF Arctic's Code of Conduct for Tour Operators relies on the efforts of the private sector to comply voluntarily. Fortunately, many tour companies throughout the Arctic actively participate in the WWF Arctic Programme. A prominent example of the successful implementation of the WWF Arctic Principles and Tour Operator Codes is by Natural Habitat Adventures, a company that for 20 years has demonstrated commitment to conservation, and consistent quality in the delivery of wildlife and nature-based tour operation experiences. Based on that record, the WWF selected the company as their sole Conservation Travel Provider. As stated by the Travel Director of WWF (Natural Habitat Adventures, 2007).

> Integral to the mission of Natural Habitat Adventures is the understanding that a complete travel experience includes protecting and preserving our natural assets and the wildlife that live in these remarkable places. Because of their dedication to this philosophy, WWF has selected Natural Habitat as their official Conservation Travel Provider to offer travel opportunities in conjunction with WWF.

Natural Habitat Adventures conducts polar tourism operations in the Arctic, Antarctic and South Georgia. As of the 2007–2008 season, they offer 15 Arctic tours with a wide range of experiences extending from wildlife viewing of polar bears, brown bears, orcas and harp seals in the Canadian and Alaskan Arctic to nature tours of Iceland and Spitsbergen, and visits to several Arctic Protected Areas. Operational good practices include:

- A philosophy of respect for the conservation of nature and local culture.
- Small groups of travelers that both minimize impacts and enhance tourist satisfaction.
- Well qualified expedition leaders that have specialized knowledge of the destinations and possess essential safety and hospitality skills. The ratio of guide to tourist is well designed to facilitate both increased appreciation of the local environment and personal safety.
- The use of the most secluded accommodations, which benefit local communities, appropriately represent native culture and reduce congestion.
- Revenue sharing with conservation organisations is an integral part of the company's philosophy and a cost knowingly paid by the tourists.
- Tourist education programs are implemented prior to travel as well as during the tour. The information contained in the programs introduce the tourist to local environments and culture, the physical conditions likely to be encountered, essential equipment and clothing, and safety considerations. All these educational efforts are intended to directly affect appropriate visitor behavior, safety and enjoyment.
- Close coordination with local governments and communities.
- Selection of most appropriate and safe modes of travel in wilderness regions. From the tourists' perspective, this minimizes potential dangers resulting from wildlife encounters, inclement weather and

natural hazards. From a resource conservation perspective, human impacts are reduced by selective routes of travel, alleviating crowding and minimal contact with the environment.

- Strong efforts are made to enlist the long-term support of the tourists for conservation programs and policies.
- Active participation in a variety of sustainable tourism programs such as the Carbon Pollution Reduction Program, the Conde Naste Traveler Green List, the Adventure Alliance, and of course WWF.

Most recently, a collection of competing tour operators, called Adventure Collection, is collaborating to promote sustainable polar tourism practices similar to those employed by Natural Habitat Adventures. It currently has 11 members that include Lindblad Expeditions, National Geographic Expeditions, Canadian Mountain Holidays, Backroads, OARS, Off the Beaten Path and Natural Habitat Adventures. Adventure Collection is also strongly affiliated with Luxury Alliance, a group of cruise companies and luxury hotel organisations.

The significance of these alliances is that they offer strong evidence that the good practices advocated by WWF Arctic Programme are gaining acceptance by the polar tourism industry.

Association of Arctic Expedition Cruise Operators (AECO) – Guidelines for Expedition Cruise Operations in the Arctic

The AECO guidelines are tools for the organization of respectable, environmentally-friendly and safe expedition cruising in the Arctic by the members. The guidelines are intended to support AECO members in their efforts to give their visitors memorable and safe experiences of the Arctic's unique and fragile nature, wildlife, cultures and cultural remains. The guidelines are also intended to support the protection of the environment and respect for and benefits to local communities.

Tourism, cruise and shipping activities in the Arctic operate within a comprehensive framework of international and national laws and regulations to ensure safety and preservation of the environment. Nevertheless, there is a need for operators to take responsibility for their activities and actions both within formal laws and regulations, and also where these regulations do not reach or define all aspects of their activities.

The expedition cruises conducted by all AECO members represent the sole means of access to the public (except for the very resourceful few) to the more remote areas of the Arctic. We believe that access to these areas should be kept open to the public, unless very strong reasons require closure of some kind. AECO believes that the best way to secure access to the tourist operators is through professional and sound organization and management. AECO members are prepared to take responsibility for their part of this management by operating according to laws and regulations, and through implementation of self-regulation.

All AECO-members already work according to a large set of operating manuals and internal guidelines, and in accordance with existing laws and regulations. The AECO- guidelines are not meant to replace member companies' operating manuals, but to supplement and strengthen the set of available management tools. AECO also appreciate that the individual member companies might focus on specific aspects of the arctic experience and environment through theme programs, and that this might put more emphasis on specific areas of for example, environmental protection, than what is specified in the AECO-guidelines.

The October 2007 Table of Contents for

AECO'S GUIDELINES FOR EXPEDITION CRUISE OPERATIONS IN THE ARCTIC

AECO RESPONSIBILITIES
LEGISLATION
- International
- Svalbard
- Jan Mayen
- Greenland

PLANNING, PREPARATION AND IMPLEMENTATION
- Planning procedures
- Preparations
- Operational preparations
- Implementation of guidelines
  - External information
  - Staff members
  - Crew members
  - Onboard information
  - Recommendations

ENVIRONMENTAL CONSIDERATIONS AND SAFETY
- Landings and shore-based activities
  - General
  - Site considerations and landing plans
  - Pre-landing information for visitors
  - Litter
- Guidelines on wildlife viewing
  - General
  - Data collection

- Walrus
  - Walrus on land
  - Walrus on the ice
  - Swimming walrus
- Seals
- Whales
- Musk oxen
- Reindeer
- Arctic foxes and wolves
- Hares
- Birds
  - General
  - Birds ashore
  - Birds on water
  - Bird cliffs and colonies
- Entanglement and stranded animals
- Wildlife found dead

Vegetation

Geology

Cairns, graffiti, signs, etc.

Cultural remains

Polar bears and firearm safety

- Polar bears
- Firearms
  - General
  - Training
  - Signal guns
  - Storage and maintenance
  - In zodiacs – transportation
  - Ashore
  - In settlements

Hazards and safety risks ashore

Zodiac operations

- General safety
  - Onboard equipment and condition
- Driver qualities
- Driving procedures
- Ice conditions
- Glacier fronts
- Icebergs
- Cliffs
- Impacts on the environment
- Passengers handling procedures and instructions
- Passengers clothing/equipment

Appendix E: Provisions of the Antarctic Treaty

The Antarctic Treaty, the foundation document to which all matters and governance are referred, makes no laws of its own. Delegates of consultative parties meet at Antarctic Treaty Consultative Meetings (ATCMs), discuss issues and agree unanimously on 'measures' and 'resolutions' that they pass to their respective governments for ratification. Measures are mandatory and, if accepted unanimously, have to be taken into each government's national legislation. Resolutions are advisory, for governments to follow but not necessarily to cover by legislation.

Visitors to Antarctica thus remain under the laws and regulations of their own countries – laws that are created to be consistent with Treaty policies. Citizens of states that are not signatories have no responsibilities under the Treaty.

The Treaty's preamble acknowledges substantial contributions to scientific knowledge resulting from international cooperation, and a conviction that continuing and developing cooperation accord with the interests of science and the progress of all mankind. Article 1 affirms that Antarctica shall be used for peaceful purposes only, prohibiting the establishment of military bases or fortifications or testing of weapons. Signatories agree to continue promoting free international scientific cooperation (Article 2), including exchanges of plans, scientific personnel and information (Article 3). The Treaty dissociates itself from questions of sovereignty (Article 4), and prohibits nuclear explosions or dumping of nuclear waste within the Antarctic area (Article 5). The Treaty area is defined as south of 60°S, including ice shelves, but excluding the high seas (Article 6). The Treaty provides for inspections of stations and operations, requiring parties to exchange advance notice of their operations (Article 7), and specifies that observers and scientists working with nations other than their own are subject only to their own national jurisdictions (Article 8).

The Treaty provides for consultative meetings of representatives to exchange information, consult together on matters of common interest and recommend to their governments measures in furtherance of the Treaty's objectives (Article 9). It requires contracting parties to ensure that its principles and purposes are observed and provides for means of resolving disputes between parties, methods of modification, amendment and review, and means of accession to the Treaty by other states (Articles 10–13). The final Article provides for copies in four languages (English, French, Russian, Spanish) to be deposited and distributed to signatory and acceding states.

Under Article 13, the Treaty is open for accession by any state that is a member of the United Nations. States that demonstrate interest in Antarctica by undertaking substantial scientific activity may become members of an administrative inner circle, the Antarctic Treaty Consultative Parties (ATCPs). There are currently (2008) 46 contracting parties, of which 28 are ATCPs.

Appendix F

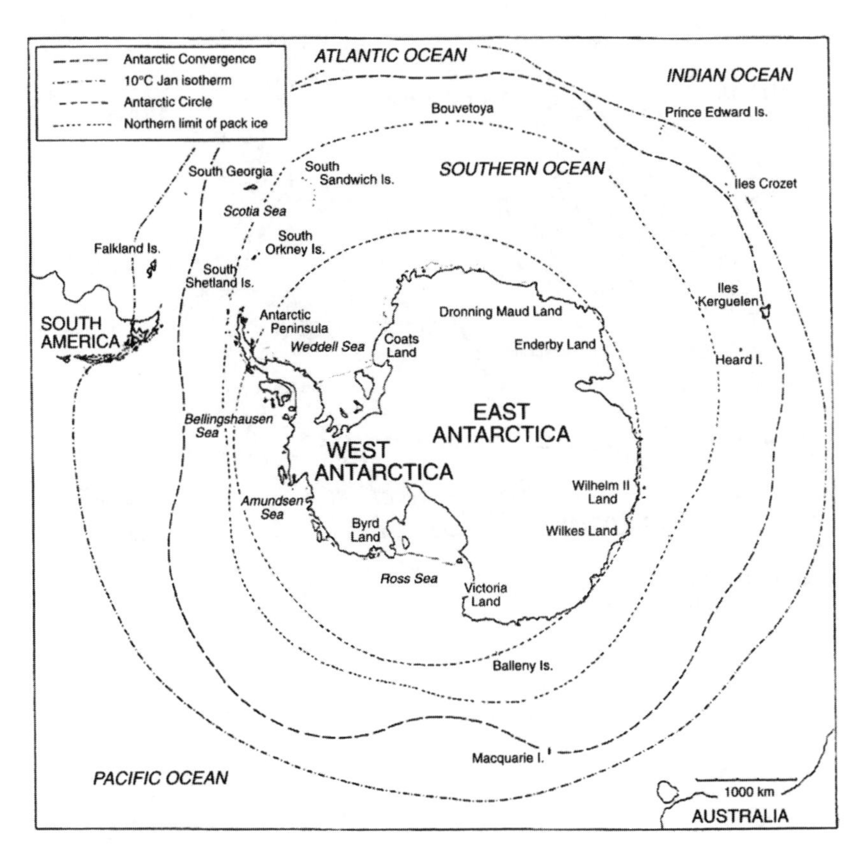

The Antarctic region
Source: Stonehouse, 1989.

The Arctic region
Source: Stonehouse, 1989.

References

ACIA (2004) *Impacts of a Warming Arctic: Arctic Climate Impact Assessment.* Cambridge: Cambridge University Press.

Alaska Native Council (1995–2005) *Alaska Native Journeys.* Anchorage, AK: Alaska Native Council.

Alberts, F.G. (1995) *Geographic Names of the Antarctic.* Arlington, VA: National Science Foundation.

Amberger, R.L. (2003) Living cultures – living parks in Alaska: Considering the reconnection of native peoples to their cultural landscapes in parks and protected areas. In A.E. Watson and J. Sproull (eds) *Science and Stewardship to Protect and Sustain Wilderness Values: Seventh World Wilderness Congress Symposium.* Proceedings. RMRS-P-27, USDA. Ogden, UT: Forest Service, Rocky Mountain Research Station.

Arctic Marine Shipping Assessment 2009 Report, Arctic Council, April 2009.

Arctic Monitoring and Assessment Programme (1998) *AMAP Assessment Report: Pollution Issues.* Oslo: AMAP.

Basberg, B. (2004) *The Shore Whaling Stations at South Georgia: A Study in Antarctic Industrial Archaeology.* Oslo: Novus Forlag.

Bauer, T. (2007) Antarctic scenic overflights. In J. Snyder and B. Stonehouse (eds) *Prospects for Polar Tourism* (pp. 188–197). Wallingford: CABI.

Beaglehole, J.C. (1974) *The Life of Captain James Cook.* London: A&C Black.

Beltramino, J.C.M. (1993) *The Structure and Dynamics of Antarctic Populations.* New York. Vantage Press.

Bertram, E. (2005) Tourists, gateway ports and the regulation of shipborne tourism in wilderness regions: The case of Antarctica. Unpublished PhD thesis, Royal Holloway, University of London.

Bertram, E. (2007) Antarctic shipborne tourism: An expanding industry. In J. Snyder and B. Stonehouse (eds) *Prospects for Polar Tourism* (pp. 149–169). Wallingford: CABI.

Bertram, E. and Stonehouse, B. (2007) Tourism management for Antarctica. In J. Snyder and B. Stonehouse (eds) *Prospects for Polar Tourism* (pp. 285–309). Wallingford: CABI.

Bertram, E., Gunn, C. and Stonehouse, B. (2008) The cruise of MS Golden Princess in Antarctic waters, January 2007. *Polar Record* 44 (229), 177–180.

Bertram, E., Muir, S. and Stonehouse, B. (2007) Gateway ports in the development of Antarctic tourism. In J. Snyder and B. Stonehouse (eds) *Prospects for Polar Tourism* (pp. 123–146). Wallingford: CABI.

Blamey, R.K. (2001) Principles of ecotourism. In D.B. Weaver (ed.) *The Encyclopaedia of Ecotourism* (pp. 5–22). Wallingford: CABI.

Blanchette, R.A., Held, B.W., Jurgens, J.A., Aislabie, J., Duncan, S. and Farrell, R.L. (2004) Environmental pollutants from the Scott and Shackleton expeditions during the 'Heroic Age' of Antarctic exploration. *Polar Record* 40 (213), 143–151.

Boczec, B.A. (1988) The legal status of visitors, including tourists, and non-governmental expedition in Antarctica. In R. Wulfrum (ed.) *Antarctic Challenge III* (pp. 455–490). Berlin: Dunker and Humbolt.

Bogen, H.S.I. (1957) *Main Events in the History of Antarctic Exploration*. Sandefjord: Norsk Hvalfangst Tidende.

Bogoyavlenskiy, D. and Siggner, A. (2004) Arctic demography. In N. Einarsson, J.N. Larsen, A. Nilsson and O.R. Young (eds) *Arctic Human Development Report* (pp. 27–41). Akureyri: Stefansson Arctic Institute.

Boo, E. (1990) *Ecotourism: The Potentials and Pitfalls* (2 vols). Washington, DC: World Wildlife Fund.

Brigham, L. and Ellis, B. (eds) (2004) *Arctic Marine Transport Workshop 28–30 September 2004*. Anchorage, AK: Institute of the North, US Arctic Research Commission.

Broadbent, N. (1992) Reclaiming US Antarctic history: The restoration of East Base, Stonington Island. *Antarctic Journal of the United States* 27 (2), 14–17.

Brodie, H. (1987) *Living Arctic: Hunters of the Canadian North*. London: Faber and Faber.

Buckley, R. (2001) Environmental impacts. In D.B. Weaver (ed.) *The Encyclopaedia of Ecotourism* (pp. 379–394). Wallingford: CABI.

Budke, I. (1999) Community-based tourism development in Nunavut, involving people and protecting places. In I. Budke and P.W. Williams (eds) *On Route to Sustainability, Best Practices in Sustainable Tourism* (pp. 39–47). Ottawa: Canadian Tourism Commission and the Centre for Tourism Policy and Research, Simon Fraser University.

Burroughs, J., Muir, J. *et al.* (1986) *Alaska: The Harriman Expedition, 1899* (p. 383). New York: Dover Publications.

Butler, R.W. (1991) Tourism, environment and sustainable development. *Environmental Conservation* 18 (3), 201–209.

Butler, R.W. (1992) Alternative tourism: The thin end of the wedge. In V.L. Smith and W.R. Eadington (eds) *Tourism Alternatives: Potentials and Problems in the Development of Tourism* (pp. 31–46). Philadelphia, PA: University of Pennsylvania Press.

Canadian Ice Service (2009) http://iceglaces.ec.gc.ca/App/WsvPageDsp.cfm?ID = 11912&Lang = eng

Canadian Press (11/20/07) Inuit-owned cruise line introduces trip to Ellesmere Island – Online document: Canadian Press, 11/20/07, http://www.cruisenorthexpeditions.com 2010

Cater, E. (1994) Introduction. In E. Cater and G. Lowman (eds) *Ecotourism a Sustainable Option?* (pp. 3–17). New York: John Wiley & Sons.

Cater, E. and Lowman, G. (1994) *Ecotourism a Sustainable Option?* New York: John Wiley & Sons.

Ceballos-Lascurain, H. (1987) The future of ecotourism. *Mexico Journal* January, 13–14.

Codling, R.J. (1982) Seaborne tourism in the Antarctic: An evaluation. *Polar Record* 21 (130), 3–10.

Codling, R.J. (1995) The precursors of tourism in the Antarctic. In C.M. Hall and M.E. Johnston (eds) *Polar Tourism: Tourism in the Arctic and Antarctic Regions* (pp. 167–177). Chichester: John Wiley & Sons.

Collis, S.M. (1890) *A Woman's Trip to Alaska, being an Account of a Voyage through the Inland Seas of the Sitkan Archipelago in 1890*. New York: Cassell Publishing Co.

Conway, W.M. (1897) *The First Crossing of Spitsbergen*. London: Dent.

Cornwallis, G., Bender, A. and Swaney, D. (2002) *Norway.* Melbourne: Lonely Planet.

Cox, S.S. (1882) *Arctic Sunbeams: Or from Broadway to the Bosphorus by way of the North Cape.* New York: G.P. Putnam's Sons.

Crosbie, K. (1998) Monitoring and management of tourist activities in the Maritime Antarctic. Unpublished PhD thesis, University of Cambridge.

Delgado, J.P. (1999) *Across the Top of the World: The Quest for the Northwest Passage.* New York: Checkmark Books.

Dent, C.T. (1892) *The Badminton Library of Sports and Pastimes – Mountaineering.* London: Longmans, Green, and Co.

Dingwall, P.R., Fraser, C., Gregory, J.G. and Robertson, C.J.R. (1999) *Enderby Settlement Diaries.* Wellington: Wild Press and Wordsell Press.

Dodds, K. (1997) *Geopolitics in Antarctica: Views from the Southern Oceanic Rim.* Chichester: John Wiley & Sons.

Douglas, W.O. (1960) *My Wilderness – The Pacific Northwest.* New York: Doubleday & Co.

Dressler, W.H., Berkes, F. and Mathias, H. (2001) Beluga hunters in a mixed economy: Managing the impacts of nature-based tourism in the Canadian western Arctic. *Polar Record* 37 (200), 35–48.

du Chaillu, P.B. (1881) *The Land of the Midnight Sun – Summer and Winter Journeys through Sweden, Norway, Lapland and Northern Finland* (2 vols). New York: Harper & Brothers.

Dufferin, L. (1873) *A Yacht Voyage Letters from High Latitudes: Being some Account of a Voyage, in 1856, in the Schooner Yacht* Foam, *to Iceland, Jan Mayen, and Spitzbergen.* New York: Lovell, Adam, Wesson & Co.

Enzenbacher, D.J. (1992) Tourists in Antarctica: Numbers and trends. *Polar Record* 28 (164), 17–22.

Enzenbacher, D.J. (1995) The regulation of Antarctic tourism. In C.M. Hall and M.E. Johnston (eds) *Polar Tourism; Tourism in the Arctic and Antarctic Regions* (pp. 179–215). Chichester: John Wiley & Sons.

Feshbach, M. and Friendly, A. (1992) *Ecocide in the USSR. Health and Nature under Siege.* New York: Harper Collins.

Ford, J.D., Smit, B., Wandel, J. and MacDonald, J. (2006) Vulnerability to climate change in Iglook, Nunavut: What can we learn from the past and present. *Polar Record* 42 (221), 127–138.

Fraser, C. (1986) *Beyond the Roaring Forties: New Zealand's Subantarctic Islands.* Wellington: Government Printing Office.

Frome, M. (1974) *Battle for the Wilderness.* New York: Praeger Publishers.

Fuchs, V.E. (1982) *Of Ice and Men.* Oswestry: Anthony Nelson.

General Accounting Office (2000 *et. seq.*) *Marine Pollution: Progress made to Reduce Marine Pollution by Cruise Ships, but Important Issues Remain.* Washington, DC: General Accounting Office.

Goldman, M.I. (1972) *Environmental Pollution in the Soviet Union: The Spoils of Progress.* Cambridge, MA: MIT Press.

Gorman, B. (2005) Future of Canadian Arctic shipping. In L. Brigham and B. Ellis (eds) *Arctic Marine Transport Workshop 28–30 September 2004* (A-6). Anchorage, AK: Institute of the North.

Greenland Tourism & Business Board (2006 *et. seq.*) *Annual Report – Greenland Port Statistics for 2006.* Greenland: Nuuk.

Griffiths and Young (1991) Sustainable development in the Arctic. In *Challenges of a Changing World: Festschrift to Willy Østreng* (pp. 33–53). Polhøgda: Fridtjof Nansen Institute.

Haase, D., Stoney, B. Mcintosh, A., Carr, A. and Gilbert, N. (2007) Stakeholder perspectives on regulatory aspects of Antarctic tourism. *Tourism in Marine Environment* 4 (2-3), 167–183.

Haase, D., Lamers, M. and Amelung, B. (2009) Heading into uncharted territory? Explosing the institutional robustness of self-regulation in the Antarctic tourism sector. *Journal of Sustainable Tourism* 17 (4), 411–430.

Haber, E. (1986) Flora of the circumpolar Arctic. In B. Sage (ed.) *The Arctic and its wildlife* (pp. 59–71). Beckenham: Croom Helm.

Harpaz, B.J. (2005) Travel forecast is sunny as 2005 begins. *The Denver Post*, 23 January, 8T.

Hart, I.B. (2001) *Pesca – A History of the Pioneer Modern Whaling Company in the Antarctic.* Salcombe: Aiden Ellis.

Hart, I.B. (2006) *Whaling in the Falkland Islands Dependencies 1904–1931.* Newton St. Margarets: Pequena.

Hattersley-Smith, G. (1991) The history of place names in the British Antarctic Territory. *British Antarctic Survey Scientific Reports* 113 (Parts 1 and 2).

Headland, R.K. (1989) *Chronological List of Antarctic Expeditions and Related Historical Events.* Cambridge: Cambridge University Press.

Headland, R.K. (2004) Northwest Passage voyages. In L. Brigham and B. Ellis (eds) *Arctic Marine Transport Workshop 28–30 September 2004* (A-8, A-20). Anchorage, AK: Institute of the North, US Arctic Research Commission.

Heap, J. (1994) *Handbook of the Antarctic Treaty System* (8th edn). Washington, DC: US Department of State.

Hemmings, A. and Roura, R. (2003) A square peg in a round hole: Fitting impact assessment under the Antarctic Environmental Protocol to Antarctic tourism. *Impact Assessment and Project Appraisal* 21 (1), 13–24.

Hendee, J., Stankey, G. and Lucas, R. (1990) *Wilderness Management* (2nd edn). Golden, CO: North American Press.

Herbert, W. (1976) *Eskimos.* London: Collins.

Hopkins, D.M. (ed.) (1967) *The Bering Land Bridge.* Stanford, CA: Stanford University Press.

Hughes, J. and Davis, B. (1995) The management of tourism at historic sites and monuments. In C.M. Hall and M.E. Johnston (eds) *Polar Tourism: Tourism in the Arctic and Antarctic Regions* (pp. 235–255). Chichester: John Wiley & Sons.

Huntington, H.P. (1992) *Wildlife Management and Subsistence Hunting in Alaska.* London: Bellhaven Press.

Huntington, H.P., Freeman, M., Lucey, B., Spearman, G. and Whiting, A. (2007) Tourism in rural Alaska. In J.M. Snyder and B. Stonehouse (eds) *Prospects for Polar Tourism* (pp. 71–83). Wallingford: CABI.

Huxley, L. (1914) *Scott's Last Expedition.* London: Smith, Elder and Co.

Icelandic Tourist Board (2006) *Your Official Guide to Iceland.* Reykjavik: Icelandic Tourist Board.

Imbrie, J. and Imbrie, K.P. (1979) *Ice Ages: Solving the Mystery.* Enslow: Shore Hills.

Issaverdis, J.-P. (2001) The pursuit of excellence: Benchmarking, accreditation best practice and auditing. In D.B. Weaver (ed.) *The Encyclopaedia of Ecotourism* (pp. 5579–5594). Wallingford: CABI.

Jensen, O. (2008) Arctic shipping guidelines: Towards a legal regime for navigation safety and environmental protection. *Polar Record* 44 (229), 107–114.

Jernsletten, J.-L. and Klokov, K.B. (2002) *Sustainable Reindeer Husbandry.* Tromsö: University of Tromsö Press.

Johnston, M.E. and Hall, C.M. (1995) Visitor management and the future of tourism in polar regions. In C.M. Hall and M.E. Johnston (eds) *Polar*

Tourism: Tourism in the Arctic and Antarctic Regions (pp. 297–313). Chichester: John Wiley & Sons.

Karr, J.R. and Dudley, D.R. (1981) Ecological perspective on water quality goals. *Environmental Management* 5, 55–56.

Kerry, K.R. and Hempel, G. (eds) (1990) *Antarctic Ecosystems: Ecological Change and Conservation.* Berlin: Springer-Verlag.

Khitun, O. and Rebristaya, O. (2002) Anthropogenic impacts on habitat, structure and species richness in the west Siberian Arctic. In A.E. Watson, L. Lalessa and J. Sproull (compilers) *Wilderness in the Circumpolar North: Searching for Compatability in Ecological, Traditional and Ecotourism Values* (p. 143). Ogden: Rocky Mountain Research Station.

Kneeland, S. (1876) *An American in Iceland. An Account of its Scenery, People and History with a Description of its Millenial Celebration in August, 1874.* Boston, MA: Lockwood, Brooks and Company.

Koh, T.T.B. (1991) Negotiating a new world order for the sea. In Fridtjof Nansen Institute (eds) *Challenges of a Changing World: Festschrift to Willy Østreng* (pp. 91–108). Polhøgda: Fridtjof Nansen Institute.

Kriwoken, L. and Rootes, D. (2000) Tourism on ice: Environmental impact assessment of Antarctic tourism. *Impact Assessment and Project Appraisal* 18 (2), 138–150.

Kriwoken, L.K., Jabour, J. and Hemmings, A.D. (eds) (2007) *Looking South: Australia's Antarctic Agenda.* Annandale: Federation Press.

Krupnik, I. (1993) *Arctic Adaptations: Native Whalers and Reindeer Herders of Northern Eurasia.* Hanover, NH: University Press of New England.

Laletin, A.P., Vladishevskii, D.V. and Vladishevskii, A.D. (2002) Protected areas of the central Siberian Arctic. In A. E. Watson, L. Lalessa and J. Sproull (compilers) *Wilderness in the Circumpolar North: Searching for Compatibility in Ecological, Traditional, and Ecotourism Values* (p. 143), 2001 May 15-16; Anchorage, AK. Proceedings. RMRS-P-26, Ogden, UT, USDA, Forest Service, Rocky Mountain Research Station.

Lamb, H.H. (1972) *Climate: Past, Present and Future. 1. Fundamentals and Climate Now.* London: Methuen.

Lamers, M., Stel, J.H. and Amelung, B. (2007) Antarctic adventure tourism and private expeditions. In J. Snyder and B. Stonehouse (eds) *Prospects for Polar Tourism* (pp. 170–187). Wallingford: CABI.

Landau, D. and Splettstoesser, J. (2007) Antarctic tourism: What are the limits? In J. Snyder and B. Stonehouse (eds) *Prospects for Polar Tourism* (pp. 197–209). Wallingford: CABI.

Lange, M.A. and Flanders, N.E. (1993) *The Effects of Global Changes on Social Life in the Arctic and Possible Response Strategies. Arctic Centre Reports No. 9.* Rovaniemi: Arctic Centre, University of Lapland.

Larsen, J.N. (ed.) (2004) *Arctic Human Development Report.* Akureyri: Steffanson Arctic Institute.

LeMasurier, W.E. (1990) Late Cenozoic volcanism on the Antarctic plate: An overview. In W.E. LeMasurier and J.W. Thomson (eds) *Volcanoes of the Antarctic Plate and Southern Oceans. Antarctic Research Series 48* (pp. 1–18). Washington, DC: American Geophysical Union.

Leopold, A. (1949) *A Sand County Almanack.* Oxford: Oxford University Press.

Levinson, J.M. and Ger, E. (1998) *Safe Passage Questioned; Medical Care and Safety for the Polar Tourist.* Centreville, MD: Cornell Maritime Press.

Lindblad, L.-E. and Fuller, J.G. (1983) *Passport to Anywhere: The Story of Lars-Erik Lindblad.* New York: New York Times Books.

Lindsey, C.C. (1981) Arctic refugia and the evolution of arctic biota. In G.G. Scudder and J.L. Reveal (eds) *Evolution Today. Proceedings of the Second International Congress of Systematic and Evolutionary Biology, Hunt Institute for Botanical Documentation* (pp. 7–10). Pittsburgh, PA: Carnegie-Mellon University.

Loomis, J.B. (1993) *Integrated Public Lands Management Principles and Applications to National Forests, Parks, Wildlife Refuges, and BLM Lands.* New York: Columbia University Press.

Lucas, R.C. (1985) Visitor use characteristics, attitudes, and use patterns in the Bob Marshall Wilderness Complex, 1970–82. *Research Paper INT-345.* Ogden, UT: US Department of Agriculture, Forest Service, Intermountain Research Station.

Mahon, P. (1984) *Verdict on Erebus.* Auckland: Collins.

Marshall, R. (1933a) *Arctic Village.* New York: The Literary Guild.

Marshall, R. (1933b) *The People's Forests.* New York: Harrison Smith and Robert Haas.

McCloskey, M.E. (1970) *Wilderness the Edge of Knowledge.* San Francisco, CA: Sierra Club.

McGhee, R. (2001) *The Arctic Voyages of Martin Conway.* London: British Museum Press.

McGrath, G. (2007) Arctic cruisers drawn by teeming wildlife. *Times Online,* 9 August.

McIntosh, E. and Walton, D.W.H. (2000) *Environmental Management Plan for South Georgia.* Cambridge: British Antarctic Survey.

McKercher, B. (1993) The unrecognised threat to tourism: Can tourism survive 'sustainability'? *Tourism Management* 14 (2), 131–136.

McKercher, B. (2001) The business of ecotourism. In D.B. Weaver (ed.) *The Encyclopaedia of Ecotourism* (pp. 565–577). Wallingford: CABI.

Mellor, M. and Swithinbank, C. (1989) *Airfields on Antarctic Glacier Ice. CRREL Report 89–21.* Washington, DC: National Science Foundation Division of Polar Programs.

Mieczkowski, Z. (1995) *Environmental Issues of Tourism and Recreation.* Lanham, MD: University Press of America.

Mill, H.R. (1905) *The Siege of the South Pole.* London: Alston Rivers.

Milne, S., Ward, S. and Wenzel, G. (1995) Linking tourism and art in Canada's eastern Arctic: The case of Cape Dorset. *Polar Record* 31 (176), 25–36.

Minbashian, J. (1997) Biological integrity: An approach to monitoring human disturbance in the Antarctic Peninsula region. Unpublished MPhil thesis, University of Cambridge.

Moe, A. (1991) Petroleum policy in the north. In Fridtjof Nansen Institute (eds) *Challenges of a Changing World: Festschrift to Willy Østreng* (pp. 209–231). Polhøgda: Fridtjof Nansen Institute.

Münzer, U. (1985) *Iceland: Volcanoes, Glaciers, Geysers.* Luzern: Atlantis.

Murray, J. (1893) *A Handbook for Travellers in Denmark, with Schleswig and Holstein and Iceland.* London: John Murray.

Natural Habitat Adventures (2008) *2009–2010 Catalogue of the World's Greatest Nature Expeditions.* Boulder, CO: Natural Habitat Adventures.

Naveen, R., de Roy, T., Jones, M. and Monteath, C. (1989) [An Antarctic Traveller's Code.] *Antarctic Century* 4 (July–October).

Nuttall, M. (2004) *Arctic Homeland: Kinship, Community and Development in Northwest Greenland.* London: Bellhaven.

Olson, M.S. (2003) Heritage tourism becoming booming business in U.S. *The New York Times*, 19 October, 8T.

Osherenko, G. and Young, O.R. (1989) *The Age of the Arctic: Hot Conflicts and Cold Realities*. Cambridge: Cambridge University Press.

Patterson, D., Easter-Pilcher, A. and Fraser, W. (2003) The effects of human activity and environmental variability on long-term changes in Adélie penguin populations at Palmer Station, Antarctica. In A.H.L. Huiskes, W.W.C. Gieskes, J. Rozema, R.M.L. Schorno, S.M. van der Vies and W.J. Wolff (eds) *Antarctic Biology in a Global Context* (pp. 301–307). Leiden: Backhuys.

Poncet, S. and Crosbie, K. (2005) *A Visitor's Guide to South Georgia*. Maidenhead: Wildguides.

Powell, S. and Jackson, A. (2007) Australian influence in the Antarctic Treaty System: An end or a means? In J. Jabour, A.D. Hemmings and L.K. Kriwoken (eds) *Looking South: Australia's Antarctic Agenda* (pp. 38–53). Sydney: Federation Press.

Reel, M. (2007) Cruise ship sinks off Antarctica. *Washington Post*, 24 November, AO1.

Reich, R.J. (1980) The development of Antarctic tourism. *Polar Record* 20 (126), 203–214.

Robbins, M. (2007) Development of tourism in Arctic Canada. In J.M. Snyder and B. Stonehouse (eds) *Prospects for Polar Tourism* (pp. 84–101). Wallingford: CABI.

Rothwell, D.R. and Scott, S.V. (2007) Flexing Australian sovereignty in Antarctica: Pushing Antarctic Treaty limits in the national interest? In L.K. Kriwoken, J. Jabour and A.D. Hemmings (eds) *Looking South: Australia's Antarctic Agenda* (pp. 7–20). Annandale: Federation Press.

Rubin, J. (2000) *Antarctica*. Melbourne: Lonely Planet.

SCAR (1998) *Composite Gazetteer of Antarctica*. Rome: Scientific Committee on Antarctic Research.

Scidmore, E.R. (1885) *Alaska: Its Southern Coast and the Sitkan Archipelago*. Boston, MA: D. Lothrop and Co.

Snyder, J. and Shackleton, K. (1986) *Ship in the Wilderness*. London: Dent and Sons.

Snyder, J.M. (2003) A Systems Approach to Sustainable Resource Management in South Georgia: Applying the Approach to Grytviken. Paper presented at The Future of South Georgia: South Georgia Association Conference, Centennial, Col., Strategic Studies Inc.

Snyder, J.M. (2007) Pioneers of polar tourism and their legacy. In J. Snyder and B. Stonehouse (eds) *Prospects for Polar Tourism* (pp. 15–31). Wallingford: CABI.

Snyder, J.M. and Stonehouse, B. (2007) *Prospects for Polar Tourism*. Wallingford: CABI.

Soper, T. (1994) *Antarctica: A Guide to the Wildlife*. Chalfont St. Peter: Bradt.

Soper, T. (2001) *The Arctic: A Guide to Coastal Wildlife*. Chalfont St. Peter: Bradt.

Spellerberg, I.F. (1991) *Monitoring Ecological Change*. Cambridge: Cambridge University Press.

Spude, C.H. and Spude, R.L. (1993) *East Base*. Denver: US Department of the Interior, National Parks Service.

Stankey, G.H., Cole, D.N., Lucas, R.C., Petersen, M.E. and Frissell, S.S. (1985) The Limits of Acceptable Change (LAC) system for wilderness planning. *General Technical Report INT-176*. Ogden, UT: US Department of Agriculture, Forest Service, Intermountain Research Station.

Starobin, Paul. June 2008. *Frozen Assets: Awash in oil wealth, Siberia goes upscale*. National Geographic Magazine (pp. 60–85), Volume 213, No. 6, Washington, DC.

State of Alaska (2008) *Alaska Visitor Survey Profile 2007.* Juneau, Alaska: Division of Tourism, State of Alaska.

Stewart, E.J., Howell, S.E.L., Draper, D., Yackel, J. and Tivy, A. (2007) Sea ice in Canada's Arctic: Implications for cruise tourism. *Arctic* 60 (4), 370–380.

Stonehouse, A.F. (2009) *Essential Iceland* (2nd edn). Basingstoke: Automobile Association Publishing.

Stonehouse, B. (ed.) (1986) *Arctic Air Pollution.* Cambridge: Cambridge University Press.

Stonehouse, B. (1990a) *North Pole, South Pole: A Guide to the Ecology and Resources of the Arctic and Antarctic.* London: Prion.

Stonehouse, B. (1990b) A traveller's code for Antarctic visitors. *Polar Record* 26 (156), 56–58.

Stonehouse, B. (1991) Polar ecosystems, management and climate modelling. In C. Harris and B. Stonehouse (eds) *Antarctica and Global Climate Change* (pp. 147–154). London: Belhaven Press.

Stonehouse, B. (1992) IAATO: An association of Antarctic tour operators. *Polar Record* 28 (167), 322–324.

Stonehouse, B. (2002) *Encyclopaedia of Antarctica and the Southern Oceans.* London: John Wiley & Sons.

Stonehouse, B. (2006) *Antarctica from South America.* Great Yarmouth: Originator Publishing.

Stonehouse, B. (2007) Tourism. In B. Riffenburgh (ed.) *Encyclopaedia of the Antarctic* (pp. 1004–1007). London: Routledge.

Stonehouse, B. and Brigham, L. (2000) The cruise of MS Rotterdam in Antarctic waters. *Polar Record* 36 (199), 347–349.

Stonehouse, B. and Crosbie, K. (1995) Tourist impacts and management in the Antarctic Peninsula area. In C.M. Hall and M.E. Johnston (eds) *Polar Tourism; Tourism in the Arctic and Antarctic Regions* (pp. 217–233). Chichester: John Wiley & Sons.

Stonehouse, B. and Snyder, J. (2007) Polar tourism in changing environments. In J. Snyder and B. Stonehouse (eds) *Prospects for Polar Tourism* (pp. 32–48). Wallingford: CABI.

Sullivan, P. (1998) Hypothermia and cold injuries; relevance to polar tourism. In J.M. Levinson and E. Ger (eds) *Safe Passage Questioned; Medical Care and Safety for the Polar Tourist* (pp. 79–95). Centreville, MD: Cornell Maritime Press.

Summers, D. (2005) *A Visitor's Guide to the Falkland Islands.* Finchley: Falklands Conservation.

Swaney, D. (1997) *Iceland, Greenland and the Faroe Islands.* Melbourne: Lonely Planet.

Swaney, D. (1999) *The Arctic.* Melbourne: Lonely Planet.

Swithinbank, C. (1997) New intercontinental air route: Cape Town to Antarctica. *Polar Record* 33 (186), 243–244.

Tracey, P. (2007) Tourism management on the southern oceanic islands. In J. Snyder and B. Stonehouse (eds) *Prospects for Polar Tourism* (pp. 32–48). Wallingford: CABI.

Triggs, G.D. (ed.) (1987) *The Antarctic Treaty Regime: Law, Environment and Resources.* Cambridge: Cambridge University Press.

Umbreit, A. (1997) *Guide to Spitsbergen.* Chalfont St. Peter: Bradt.

UN World Commission on Environment and Development (1987) *Our Common Future* (Brundtland Report). Oxford: Oxford Paperback Reference.

United States Fish and Wildlife Service (2007) *2006 National Survey of Fishing, Hunting, and Wildlife Associated Recreation*. Washington, DC: US Department of the Interior.

United States National Ice Center (2007) *Proceedings: Impact of an Ice-diminishing Arctic on Naval and Maritime Operations, July 10–12*. Washington, DC: United States National Ice Center and Arctic Research Commission.

Vidas, D. (ed.) (2000) *Protecting the Polar Marine Environment: Law and Policy for Pollution Prevention*. Cambridge: Cambridge University Press.

Viken, A. and Jorgensen, F. (1998) Tourism in Svalbard. *Polar Record* 34 (189), 123–128.

Vitebsky, P. (2006) *The Reindeer People: Living with Animals and Spirits in Siberia*. Boston, MA: Houghton Mifflin.

Walle, A.H. (1993) Tourism and traditional people: Forging equitable strategies. *Journal of Travel Research* Winter, 14–18.

Walton, D.W., Scarponi, G. and Cescon, P. (2001) A scientific framework for environmental monitoring in Antarctica. In S. Caroli, P. Cescon and D.W. Walton (eds) *Environmental Contamination in Antarctica: A Challenge in Analytical Chemistry* (pp. 33–53). Oxford: Elsevier Science.

Warren, P. (1989) Proposal for the designation and protection of Antarctic historic resources. Unpublished MA thesis, University of Seattle.

Weaver, D.B. (ed.) (2001) *The Encyclopaedia of Ecotourism*. Wallingford: CABI.

Wellman, J.D. (1987) *Wildland Recreation Policy*. New York: John Wiley & Sons.

Wight, P.A. (1994) Environmentally responsible marketing of tourism. In E. Cater and G. Lowman (eds) *Ecotourism: A Sustainable Option?* (pp. 39–56). Brisbane: John Wiley & Sons.

Williams, W.M. (1859) *Through Norway with a Knapsack*. London: Smith, Elder and Co.

World Meteorological Organization (1971) *Climatic Normals for CLIMAT and CLIMAT Ship Stations for the Period 1931–60*. Geneva: WMO.

Wright, V., Glaspell, B. and Puttkammer, A. (2001) *Linking Wilderness Research and Management: Volume 2 – Defining, Managing, and Monitoring Wilderness Visitor Experiences: An Annotated Bibliography*. Gen. Tech. Rep. RMS-GTR-79-vol. 2. Ft. Collins, CO: US Department of Agriculture, Forest Service.

Young, E. (1991) Critical ecosystems and nature conservation in Antarctica. In C. Harris and B. Stonehouse (eds) *Antarctica and Global Climate Change* (pp. 117–146). London: Belhaven Press.

Young, O.R. and Einarsson, N.E. (2004) Introduction. In J.N. Larsen (ed.) *Arctic Human Development Report* (pp. 15–26). Akureyri: Steffanson Arctic Institute.

Zelikman, L. (1989) A large scale ecological disaster is threatening from the north. *Environmental Policy Review* 3 (2), 1–8.